# The Friar and the Cipher

ALSO BY LAWRENCE AND NANCY GOLDSTONE

*Out of the Flames: The Remarkable Story of a Fearless Scholar, a Fatal Heresy, and One of the Rarest Books in the World*

*Used and Rare: Travels in the Book World*

*Slightly Chipped: Footnotes in Booklore*

*Warmly Inscribed: The New England Forger and Other Book Tales*

ALSO BY LAWRENCE GOLDSTONE

*Rights*

*Off-Line*

ALSO BY NANCY GOLDSTONE

*Trading Up: Surviving Success as a Woman Trader on Wall Street*

*Bad Business*

*Mommy and the Murder*

*Mommy and the Money*

Broadway Books
NEW YORK

# The Friar and the Cipher

ROGER BACON AND
THE UNSOLVED MYSTERY OF
THE MOST UNUSUAL MANUSCRIPT
IN THE WORLD

. . .

LAWRENCE AND NANCY
GOLDSTONE

BROADWAY



A hardcover edition of this book was originally published in 2005 by Doubleday. It is here reprinted by arrangement with Doubleday.

PRINTED IN THE UNITED STATES OF AMERICA



Visit our Web site at www.broadwaybooks.com

*Book design by Gretchen Achilles*

The Library of Congress cataloged the hardcover edition as follows:
Goldstone, Lawrence, 1947–
The friar and the cipher : Roger Bacon and the unsolved mystery of the most unusual manuscript in the world / Lawrence and Nancy Goldstone.—1st ed.
p. cm.
Includes bibliographical references and index.
1. Voynich manuscript. I. Goldstone, Nancy Bazelon. II. Title.
Z105.5.V.65G65 2004
091—dc22 2004050164

ISBN 0-7679-1472-4

1 3 5 7 9 10 8 6 4 2

TO EMILY

AND TO JED

# CONTENTS

• • •

EIGHT PAGES OF FULL-COLOR ILLUSTRATIONS FROM THE VOYNICH MANUSCRIPT ARE LOCATED AT THE CENTER OF THE BOOK

*Sad it is to think of what this great man might have given to the world . . . He held the key to treasures which would have freed mankind from ages of error and misery. With his discoveries as a basis, with his method as a guide, what might not the world have gained! Thousands of precious lives shall be lost, tens of thousands shall suffer discomfort, privation, sickness, poverty, ignorance, for lack of discoveries and methods which . . . would now be blessing the earth.*

**ANDREW DICKSON WHITE**
*Cofounder, Cornell University, 1895*

*Man, in so far as he is man, has two things, bodily strength and virtues, and in these he can be forced in many things; but he has also strength and virtues of soul. In these he can be neither led nor forced, but only hindered. And so, if a thousand times he is thrown into prison, never can he go against his will unless the will succumbs.*

**ROGER BACON**
Opus Tertium, *1268*

# Prologue

. . .

IN LATE FALL OF 1912, Wilfrid Michael Voynich, a prominent London book dealer on a buying expedition to the continent, chanced upon a collection of rare illuminated manuscripts stuffed into some old wooden chests at Villa Mondragone, a secluded castle in Frascati, Italy. The castle was then the home of a Jesuit school, but the collection had evidently been hastily secreted a century earlier when Napoléon's voracious armies were hurling Europe into war and chaos. Many of the manuscripts were marked as belonging to Pierre-Jean Beckx, the 22nd general of the Society of Jesus, who had died twenty-five years before. They also held the seals of the dukes of Ferarra, Parma, and Modena, some of Italy's preeminent noble families.

The school was badly in need of funds, and Voynich knew that he could have his pick of the lot. Aware that there was a strong market for illuminated manuscripts back in England, as they combined the lure of antiquity with rich decorations in brilliant blues, reds, and golds, he pored eagerly through the chests, sorting out the most valuable and those in the best condition. There was one

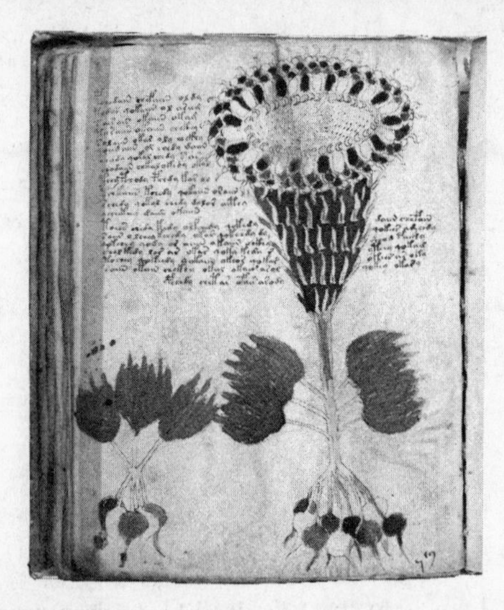

***A plant illustration from the Voynich manuscript*** BEINECKE RARE BOOK AND MANUSCRIPT LIBRARY, YALE UNIVERSITY

volume, however, that intrigued him precisely because of its lack of beauty. Voynich himself described it as an "ugly duckling." It was just over two hundred pages and small, only six by nine inches, about the size of a current-day hardcover. The pages were vellum and the cover was blank. No title or author's name was visible anywhere. There was a letter attached, on which was written a date—1665 or 1666, Voynich couldn't be sure—but he knew immediately from looking at the script and the style of the writing that the manuscript itself was actually much older.

What attracted his attention even more than the text were the illustrations that appeared on almost every page—sometimes in the margins, sometimes in the body of the work. There were hundreds of them—exotic-looking plants, quaint astronomical and astrological symbols and charts, and, most bizarre, strings of tiny naked women cavorting in a variety of fountains, waterfalls, and

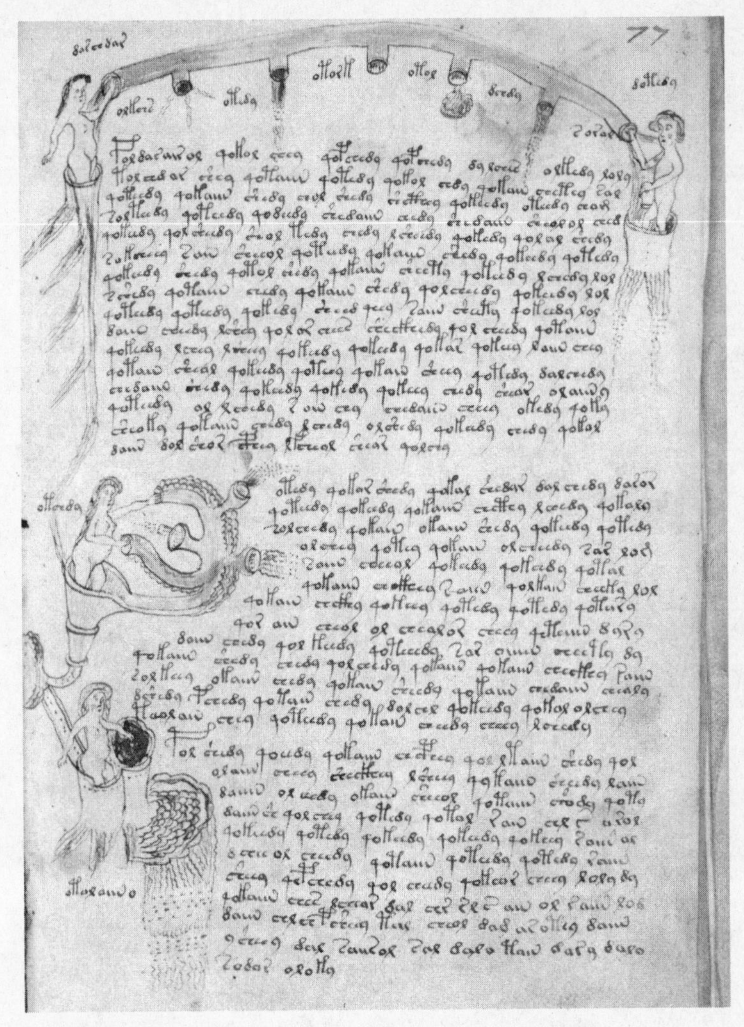

***An example of Voynich script*** **BEINECKE RARE BOOK AND MANUSCRIPT LIBRARY, YALE UNIVERSITY**

pools. The work appeared to be some sort of encyclopedia of natural science, possibly for use in the practice of medicine. Examining the vellum pages, the calligraphy, the drawings, and the pigments used, Voynich, an expert medievalist, dated the manuscript as late thirteenth century. However, when he perused the

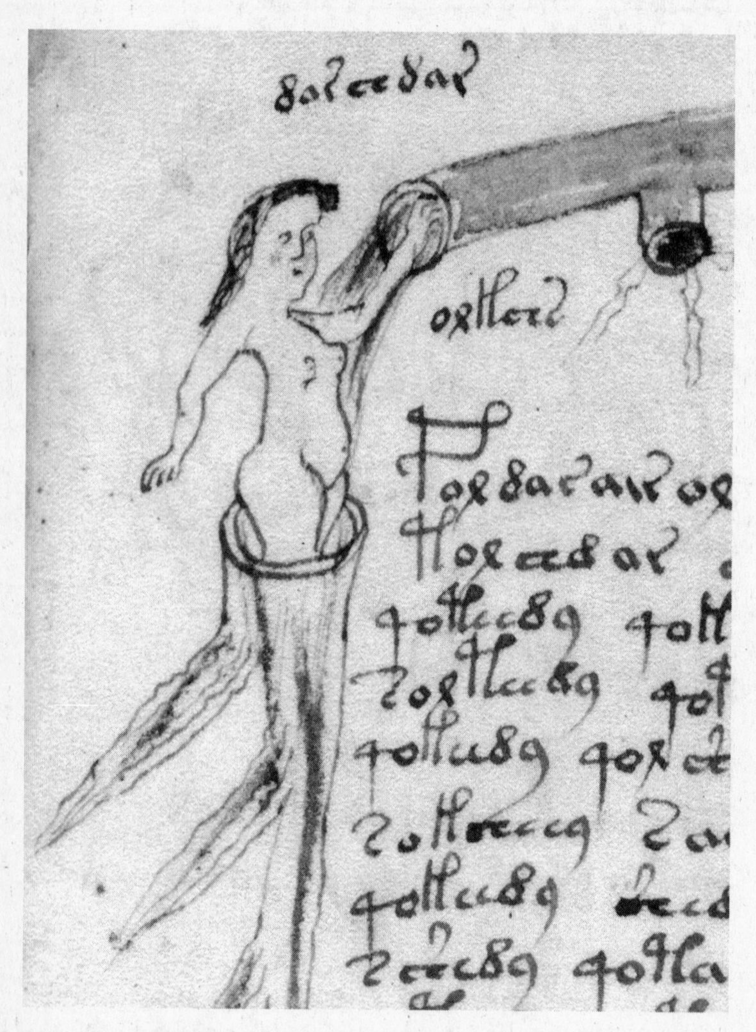

***Detail from a Voynich manuscript page*** BEINECKE RARE BOOK AND MANUSCRIPT LIBRARY, YALE UNIVERSITY

odd cursive script itself, he realized that it wasn't Latin, or any other known language for that matter. He could make no sense at all of what was written. With some further examination, he discovered why—the entire manuscript was in cipher.

No thirteenth-century manuscript of this length and detail written in cipher had ever been found before, and Voynich knew that he had stumbled upon something of exceptional importance. The sense of intrigue fascinated him as well. The knowledge had evidently been meant to be hidden. But why? What was there in these drawings of bathing women and lush plants that posed a threat to the author or to the intended reader? Or had the author taken these precautions because he feared repercussions if his book fell into the wrong hands?

Wilfrid Voynich was no stranger to conspiracy. He had been born in Lithuania in 1865, the son of Polish aristocracy. He attended the universities of Warsaw and Moscow, where he fell in with the Russian anarchist Sergius Stepniak. Stepniak, who was subsequently arrested and imprisoned for attempting to assassinate the chief of the Russian secret police, eventually escaped and fled to England. Stepniak's anarchist movement had a great influence on Lenin's older brother Alexandr, who was caught in a similar plot to assassinate the tsar and was executed in 1888.

Voynich was himself arrested by the secret police in 1885 for plotting against the Russian domination of eastern Poland. Only twenty years old, he was placed in solitary confinement in a Warsaw jail cell so cold and cramped that he developed tuberculosis and became permanently disfigured, with one shoulder always held higher than the other. After a year he escaped, but he stayed on to fight to expel the Russians. He was wounded several times before being arrested again; this time he was sent to Siberia. "Here

*Wilfrid M. Voynich, ca. 1885* COURTESY RAFAL PRINKE

I have sword, here I have sword, here I have bullet," he later told one of his rare book employees, pulling up his shirt and showing off his scars.

He escaped from Siberia as well, then, wanted and penniless, slipped across the continent and made it all the way to London, where he turned up on Stepniak's doorstep. According to the legend that grew up around him, Voynich borrowed a half crown from Stepniak and shortly thereafter appeared at the British Museum with a number of rare and unusual books that just happened to be among those that the museum was then seeking to acquire. The museum enthusiastically bought the books for an undisclosed sum, one sufficient to allow Voynich to set up shop as a rare book dealer. It also gave him the wherewithal to marry Stepniak's beautiful blond assistant, Ethel. Ethel was the youngest daughter of

George Boole, the famous Cambridge professor of mathematics who developed a system of algebraic logic that is now considered one of the precursors of computer science.

(Ethel was no stranger to conspiracy either. After marrying Voynich, she carried on a long-running affair with the World War I master spy Sidney Reilly, whom she then used as the model for the protagonist in her novel *The Gadfly*, which sold over 2.5 million copies in translation and, because of its glamorous radical hero, became one of the most beloved novels in Soviet Russia.)

Soon, the immigrant Voynich, who spoke English in a series of malapropisms, became an established member of one of Britain's most exclusive and snobbiest clubs—antiquarian book dealers. In addition to his shop on Shaftesbury Avenue, Voynich opened branches in Paris, Florence, and Warsaw. He developed a clientele that included such famous collectors as Sir William Osler and the Earl of Crawford and Balcarres, but Voynich's visitors weren't all aristocrats and proper English gentlemen. A number of shady characters of murky Eastern European origin floated in and out of his shop. One time, a Polish woman who claimed to have been horribly persecuted by the Russians, barely escaping with her life, showed up in London, seeking to enter Voynich's circle. She didn't last long. When asked by one of his employees what had happened to her—Voynich was known to support these people in perpetuity—he replied coldly, "By mistake they sent me skunk."

VOYNICH BROUGHT HIS UGLY DUCKLING MANUSCRIPT back with him to London but did not put it up for sale. Instead, he sent photographed copies of some of the pages to about twenty of the most noted cryptographers in the world, including members of MI-8, the United States Army cryptographic unit.

At first it appeared that deciphering the manuscript would be easy. With so many plants and astronomical symbols encaptioned throughout, it was simply a matter of inquiring from botanists and astronomers what the actual names were and then plugging in plaintext equivalents into what was certainly a rudimentary alphabet-substitution code. One cryptologist predicted that the manuscript would be easier to decipher than a newspaper cryptogram.

It didn't seem to work, however. Many of the plant drawings, it turned out, had no counterparts in the real world, and those that did seem identifiable did not respond to alphabet substitution. The astronomical drawings were no more enlightening. In fact, the more the cryptologists looked, the more obscure the cipher became. For six years, the manuscript resisted all attempts to make any sense of it. Even the MI-8 team, which had enjoyed brilliant success against the best German military codes in World War I, made no progress. That a thirteenth-century manuscript could confound modern code breakers so thoroughly only made the challenge greater. The Voynich manuscript was fast becoming one of the world's most famous enigmas.

Then, in 1919, William Romaine Newbold, a distinguished professor of philosophy and dean of the graduate school at the University of Pennsylvania, announced a breakthrough. Newbold, an expert in philology and cryptanalysis, noticed that on the last page, the three lines of text were written in Latin and not the symbol alphabet of the remainder of the manuscript. Working with recognizable characters, he conducted an analysis revealing that those three lines were in a different cipher—part alphabet substitution, part anagram—that was more common and over the years had become familiar to medieval scholars. After Newbold decoded what he believed was the key—a set of "instructions" for decrypting the body of the text—a further examination of the page showed that someone else had done so as well. There was a faded, almost invisible notation in one of the margins giving exactly the same solution. It was dated 1630.

In the meantime, Voynich had set himself to uncover the provenance of his discovery, hoping by this route to come to the author. When he had first dated the cipher manuscript as thirteenth century and noticed from the drawings the breadth of subjects that it apparently covered, he had decided that only two men had the scientific expertise necessary to compose such a work. One was the German Albertus Magnus (Albert the Great). Renowned for his vast contributions to theology and science, he was called *Doctor Universalis* for his encyclopedic knowledge in every area of medieval scholarship and was especially known for his role in incorporating Greek learning into the academic curriculum. He was thirteenth-century Europe's most accomplished natural scientist. As great as Albertus Magnus's achievements, however, Voynich personally favored the second man, whom he considered an "infinitely greater scholar."

This second man, dubbed *Doctor Mirabilus* (the miraculous doctor), was an expert in mathematics, astronomy, optics, al-

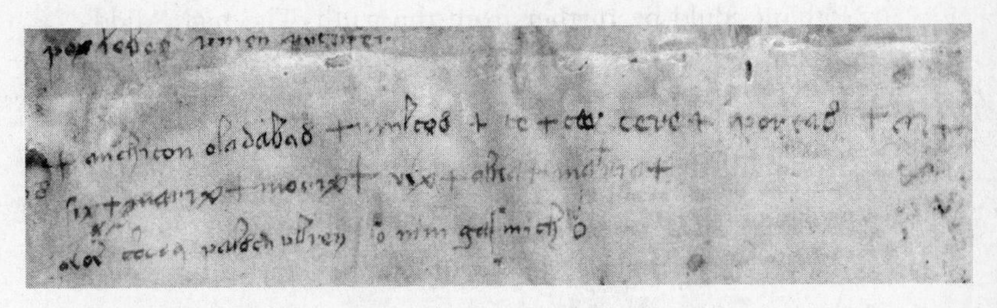

***Voynich manuscript "key." Note that characters differ from those in the body of the text.*** **BEINECKE RARE BOOK AND MANUSCRIPT LIBRARY, YALE UNIVERSITY**

chemy, languages, and homeopathic medicine. He had described the workings of the telescope and microscope four hundred years before Newton. He believed that the earth was spherical and that one could sail around it, an argument that was purported to have inspired Columbus two hundred years later. He believed that light moved at a distinct speed at a time when it was assumed that the movement was instantaneous. He questioned Galen, the great Roman anatomist and physician, and theorized about illness, disease, and the human body centuries before anatomy and medicine poked their heads into the modern age. He has sometimes been credited with inventing eyeglasses. He wrote of flying machines, motorized ships, horseless carriages, and submarines. He was the first man in Europe to describe in detail the formula for making gunpowder.

OUR MOST PREVALENT MENTAL IMAGE of the thirteenth century is of knights in chain-mail hoods with red crosses on their chests, plodding on horseback through dank forests, stopping every now and then to joust with one another in a tournament on their way to Jerusalem to scuffle futilely with the infidel Muslims. It is a dreary picture, all monks and relics, serfs and saints, poverty, piety, barbarity, and ignorance.

Nothing could be further from the truth. The high Middle Ages was a time of color, romance, scheming, and intrigue. Land, power, position, and wealth were all up for grabs, and the rule of law accommodated itself to its surroundings. It was also a period of consolidation—politically, geographically, and theologically. Monarchs moved large numbers of troops from one end of Europe to the other and even into North Africa. Sea routes were open to those who could pay for a ship. Commercial travel improved as well. It has been said that the widening of roads in Europe to accommodate oxcarts after the Dark Ages was the most significant technological advance in history.

The thirteenth century was also one of the most pivotal and exciting in the history of human knowledge. Propelled by new translations of Greek classics and the work of brilliant Islamic scholars, the best minds in Christian Europe were beginning to theorize about the power of natural science. Some had even begun to question, for the first time in a millennium, whether knowledge must be restricted to the revealed word of the Bible or could instead be gained by drawing empirical hypotheses and then testing them by experiment.

After centuries of accepting official Church interpretations as to the proper exercise of Christianity, pious Christians throughout Europe had also begun to question the opulence with which popes, cardinals, bishops, and even priests lived, often among a flock that was poor, hungry, and oppressed. How could Christ have meant for disciples to live in luxury while his more humble followers starved? Thousands joined new religious movements. The thirst to throw off the controls on what to think and what to feel offered the promise of a new enlightened age. Both science and Christianity might be forever changed.

But the medieval Church had weapons of its own. On one extreme, there was sword, fire, and fear, all of which were wielded mercilessly. Many of those who raised questions were branded as heretics, rounded up, and exterminated. The Inquisition was begun in the thirteenth century as a means to institutionalize heretical repression. But there was also piety and simple decency as many in the new Orders tried to show the world the meaning of faith and true Christian values.

Yet for all that, it seemed that the ravenous need to advance knowledge, to satisfy curiosity, *to find out,* was about to overpower Christian tradition. The battle between these two forces would shape the course of Western history for the next three hundred years, and it continues to resonate today. Each side in that struggle had as its advocate a great scientific philosopher. Each was deeply pious, each believed in the total supremacy of God. One was born in England in 1214, the other in Italy a decade later. One attempted to add natural science to theology, the other to make theology a natural science. One was imprisoned, one dined with kings. One died anathemized, the other became a saint. The Italian was Thomas Aquinas, the protégé of Albertus Magnus. The Englishman was the very man who Voynich believed had authored his cipher manuscript.

Now William Romaine Newbold at the University of Pennsylvania seemed to have confirmed Voynich's guess, for the first line of the key when deciphered read:

"To me, Roger Bacon."

FROM THE DAY THAT NEWBOLD ANNOUNCED that he had deciphered the key, the name and reputation of Roger Bacon has been inextricably linked to the Voynich manuscript. The mystery is almost irresistible—did the brilliant medieval scientist, an expert in secret languages and codes, hounded by reactionaries in his own time, compose what is arguably the most enigmatic artifact ever uncovered? If he did write it, what does it say? Does it hide a great scientific discovery for which Bacon did not get credit? Who was he writing it for? And if he did not draft the manuscript, who did, and why is Bacon's name on it?

To make some sense of the puzzle, it is necessary to study the man as well as the manuscript. In its own way, the life of Roger Bacon is as much a cipher as the sheaf of papers Wilfrid Voynich pulled from the trunk in the Villa Mondragone. Many of the details of his life remain shadowy, and the mythology that grew up around him often blurred fact with fiction. Much of what we know has been deduced from background information rather than gleaned from irrefutable evidence.

Still, as Bacon has been a subject of almost cultish fascination for more than four centuries, a remarkable amount of scholarly detective work has been done to lift his biography from the murk. Bacon has aroused considerable passions from both those who think that he was one of the preeminent figures in the history of science, on a par with Galileo, and those who dismiss him as a quirky iconoclast whose appeal is derived from his quixotic personality rather than any real contribution to human thought. In order to objectively navigate between these extremes, a historian must rely in no small part on what Bacon's detractors are forced to admit and what his champions are forced to concede.

The manuscript's provenance is equally obscure. There are large gaps in the chain of ownership. A trail must be woven, once again not by hard data but by careful construction, the testing of relative hypotheses, the weighing of alternatives.

The scale of the investigation is breathtaking; it covers the range of human experience, skipping across time periods and disciplines. It begins with the musings of Plato and Aristotle, moves through the development of scientific and Christian thought on to the rise of universities and finally to the intellectual standoff between Roger Bacon and his belief in experiment and Thomas Aquinas and the logic of Catholicism. But Bacon's legacy and the foggy history of the manuscript also lead to spies and plots in Elizabethan England, palaces in exotic Bohemia, a spectacular showman and scientist who in his lifetime was considered a rival to Isaac Newton, and finally to modern cryptanalysis and the secret world of the premier code-breaking unit in America, the National Security Agency.

It is a journey of intellectual curiosity, moral courage, and tenacity of spirit, which is incomplete even to this day. Still, as Roger Bacon knew eight centuries ago, it is the willingness to undertake the journey, and not the results, that is so vital to human progess.

***First page of the Voynich manuscript. Under ultraviolet light, the name of a previous owner and date can be seen in the margins. Under magnification, "Tepen" can just be made out.*** BEINECKE RARE BOOK AND MANUSCRIPT LIBRARY, YALE UNIVERSITY

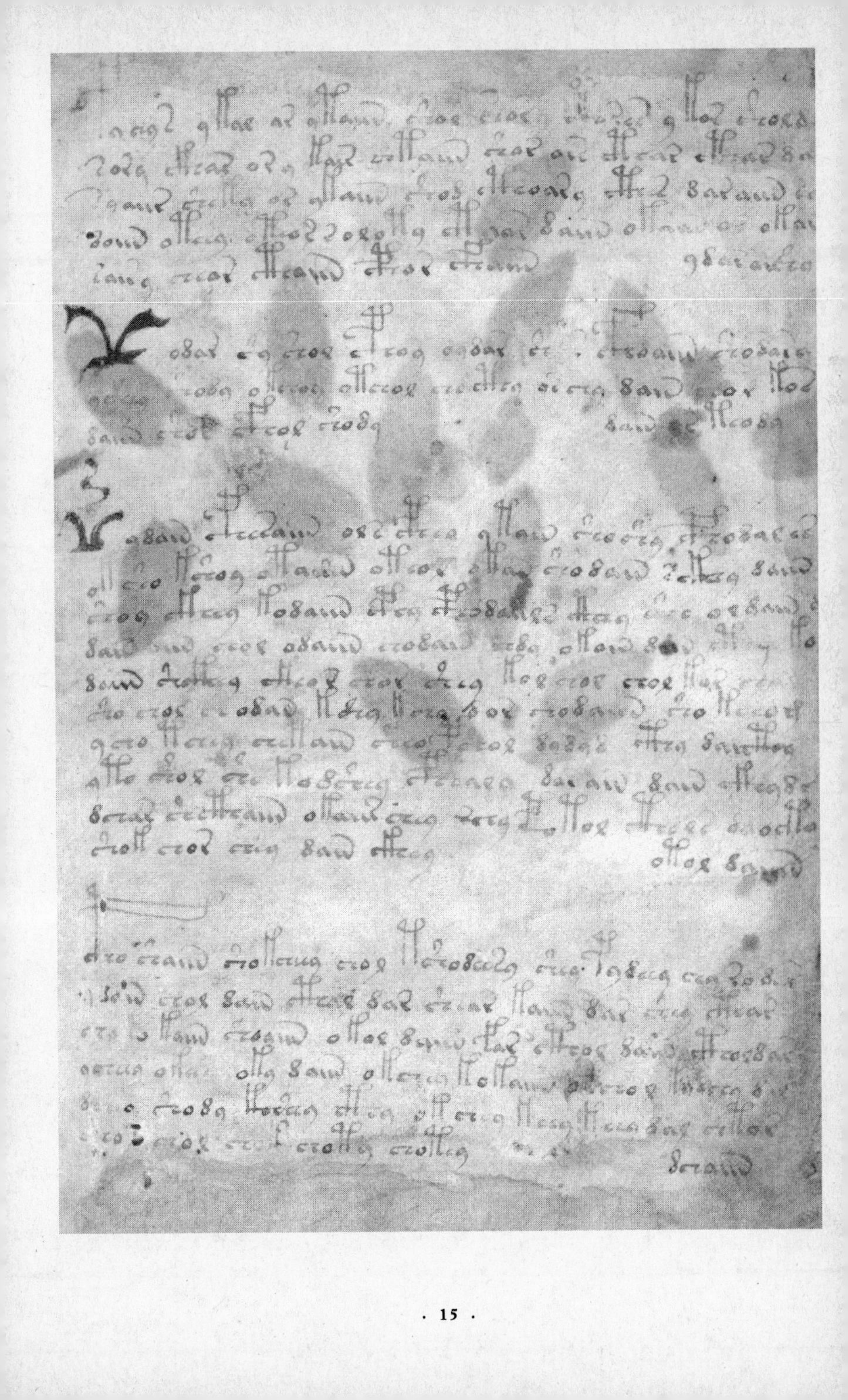

CHAPTER ONE

# Turmoil and Opportunity: Roger Bacon's England

• • •

ROGER BACON WAS BORN IN SOMERSET, in southwest England, about one hundred miles west of London. There are no surviving records of his birth—the evidence for the date comes from Bacon himself. In a work known to have been written in 1268 he said: "I have labored much in sciences and languages, and I have up to now devoted forty years to them." What he apparently meant by this was that he had started what today would be the equivalent of an undergraduate arts course in 1228. Since the average thirteenth-century boy started college at about fourteen, this puts the year of his birth at 1214. He lived to be eighty, so his lifetime spanned nearly the whole of the thirteenth century.

Bacon came from a family of wealthy minor nobles. His father held no title and was probably a product of the new and burgeoning merchant class, men who worked their way into higher society by accumulating cash, which was then used to purchase land and a manor house. The most successful of these could buy castles and conduct themselves as genuine nobility, knighting their sons, but Bacon's family did not seem to fall into this category. He had at least one older brother, to whom he refers in his writings, but neither was ever granted a title by the king.

Bacon remained throughout his life a product of the England of his childhood, an England in the midst of great change and rife with civil unrest that would soon erupt into full-scale war. The year after Bacon was born, the hapless King John was forced to sign Magna Carta and thus introduce the first glimmer of representative government into Europe. It was the very weakness of John and, later, his son Henry that created a vacuum into which political, social, educational, and, most significantly, scientific innovation rushed in. The most basic assumptions were challenged, the most fundamental truths rejected. So unfortunate was John as a ruler that he did not need to be known as John I, as no other king in the ensuing eight hundred years of English history was ever given the same name.

John was the fourth son of the tall, intense, mercurial Henry II, under whose lusty hand the kingdom had grown to encompass not only England but most of France—Normandy, Brittany, Anjou, Touraine, Toulouse, and, with his marriage to the vivacious, wily Eleanor, the Aquitaine on the Atlantic coast. The official kingdom of France, on the other hand, was limited to Paris and its environs.

Henry's two eldest sons, Henry and Richard (called Lionheart for his military prowess) were also tall and physically imposing. John was short and unattractive. Richard still referred to John as a child when he was well into his twenties. There was a third older brother, Geoffrey, who was much cleverer than John, although this in itself was not particularly noteworthy.

With all those older sons, Henry assumed that John was never going to see the English throne, so when the boy was nineteen, he tried to get him a kingdom of his own by sending him off to conquer Ireland. John left with lots of friends, three hundred mercenaries, several barrels of silver pennies with which to pay them, and the promise of a fancy gold crown fitted out with peacock feathers when he won. In no time at all, John and his friends had spent all of the pennies on themselves, causing the mercenaries to desert. He so alienated the Irish nobility that, in a place known for internecine warfare, John managed to get all the aristocrats in Ireland to band together and agree to reject him. Richard, by contrast, had subdued the powerful rebellious barons of southern France by the age of fifteen.

Richard eventually became king (Henry and Geoffrey died young) but left almost immediately on crusade, where he was captured and held for ransom by the Holy Roman Emperor. As everyone who has ever seen *The Adventures of Robin Hood* knows, during his absence, John attempted to usurp the English throne by treachery. (In truth, it was Eleanor, not Errol Flynn, who stopped him.) When John heard that his older brother had been freed and was on his way home, he turned tail and headed for France. John was so insignificant in Richard's mind that Richard forgave him and let him come home.

Just a short time later, however, Richard died while staking out a minor castle for siege. He had disdained armor while parading around the periphery and was shot in the neck with an arrow. The boy whom his father had nicknamed "Lackland" for want of a realm was crowned King John of England at Westminster Abbey on May 25, 1199. Within five years of becoming king, he had lost most of his father's French possessions to the French king, Philip Augustus, earning him a new nickname, "Softsword," among his own nobility.

Losing to the French turned out to be just the preliminary. In 1205, John, by virtue of an extremely dubious royal edict, found himself taking on the great Pope Innocent III. It was not really a fair fight.

INNOCENT III WAS ONE OF THE SEMINAL FIGURES in the history of the Church. Born into an ancient aristocratic Roman family, he proved a brilliant student in both law and theology. He enjoyed a meteoric rise through the curia and was elevated to the Throne of St. Peter while still in his thirties. Innocent inherited an institution in disarray. In the century preceding his reign, the papacy had sunk to an object of ridicule, ignored by secular monarchs. One of his predecessors had been compelled to ride backward on an ass through the countryside, and another had been obliged to flee Rome disguised as a pilgrim. By a combination of force of personality and the threat of withholding sacraments, Innocent almost single-handedly turned Rome from the political nonentity that it had become into a potent pseudostate, the most important political power in Europe. "By me kings reign and princes decree justice," he observed.

In 1205, the archbishop of Canterbury died. In Henry II's time, the English bishops would "elect" a new archbishop, although this was in fact a royal appointee. (As usual, Henry had taken custom one step further—he had not only assumed the right to elect his own man, Thomas à Becket, as archbishop, but had assumed the right to have him killed as well when Becket disagreed with him.) John naturally expected to have the same privilege as his father, but he bungled the election and Innocent claimed the right of appointment for himself. He chose the highly qualified Stephen Langton, who was at the time teaching at the University of Paris.

John was outraged at this attempt to usurp his authority and refused to let the new appointee enter the country. So began a war of wills between king and pope. Innocent placed England under interdict, which meant that English priests were forbidden to perform any of the sacraments. Suddenly, no one in England could get married, buried, or baptized. John retaliated by seizing the property of those priests who obeyed Innocent's order. To get it back, they had to swear loyalty to the crown and pay a hefty fee. They even had to pay to get their "housekeepers" (read, mistresses) back. Innocent countered by excommunicating John. The English high clergy packed up and headed to France, and by 1212 there was only one bishop left in all of England.

Still John refused to yield, so Innocent sent a message to Philip Augustus, the king of France. If Rome deposed John, the excommunicate, would Philip Augustus like to take over in his place? Philip Augustus did, in fact, want to take over England. Langton, who still had not gained entry to the country of which he was nominally the archbishop, was given letters from Innocent announcing that John had been deposed in favor of Philip Augustus. The French massed an army at the Channel.

John gave up. Langton was accepted as archbishop of Canterbury, and all of the English priests who had fled during the interdict and excommunication were restored to their property and compensated for their damages.

Philip Augustus, however, had not given up. He had not recalled his army, which was still sitting across the Channel, waiting for the order to invade. John, needing a powerful ally, turned to Innocent. To save himself, John proposed what to many in England was the unthinkable—he offered England and Ireland as fiefs of the Church, which also required that he pay a sizable monetary tribute to Rome.

France attacked anyway and was repulsed when John's subjects, to his and probably their astonishment, rallied to his aid. John, overestimating his position and their loyalty, immediately launched a counterinvasion to reclaim his father's lost territories. His armies were routed.

The English barons had had enough. A group of them took over London and forced John to put his seal to Magna Carta, or "Great Charter," originally just a remedy for a list of baronial grievances. The key provision, however, which insisted that John obtain approval on matters of state by a board of directors composed of twenty-five of his most rebellious barons, became, albeit unintentionally, a forerunner of English representative government.

John signed under duress, but he immediately sent messengers to Innocent. Innocent, who was not the type of man to encourage the spreading of power, particularly to twenty-five men who might not do what he said, wrote a strong letter threatening to excommunicate any baron who went against the king and declaring Magna Carta "null and void of all validity forever."

Uneasy truce then turned into civil war. The rebel barons responded by offering the crown of England to Philip Augustus's son Louis VIII in 1215. Louis came over with an army and secured London. He held the Channel and the east coast of England through the aid of a swashbuckling English pirate with the beguiling name of Eustace the Monk. By 1216, two-thirds of the English barons had come over to Louis. When Innocent died in July of that year, it seemed that John was finished.

And he was. Three months later, after losing all of his baggage, including his crown, in an ill-fated attempt to cross a river with his army, John consoled himself with "gluttonous consumption of peaches and new cider." He caught dysentery and died, leaving the whole mess to his nine-year-old son, Henry. By the time of Henry's coronation, held in Gloucester, the rebel barons and Louis held the north and east of England, including London, Cambridge, and York, and, thanks to Eustace the Monk, all of the coastal ports in the east except Dover.

As it turned out, however, Henry's youth was his greatest asset. The sight of this lonely boy crowned in this out-of-the-way spot without the usual pomp and ceremony made Louis appear as someone taking unfair advantage of a child. The English of the south and west rallied. Henry's regents remained loyal and, in a clever move, reissued a revised Magna Carta in Henry's name, called the Charter of Liberties, thus removing the rebel barons' original grievances (and further legitimizing limited representative government). There were defections back to the crown. A battle at Lincoln, won convincingly by the royalists, siphoned off even more of Louis's support. The decisive blow came when the royal fleet sailed windward of Eustace the Monk and, in a daring maneuver, threw powdered lime into the faces of the rebel officers, incapacitating the enemy ships. The Monk's ship was boarded and he was beheaded, and it was all over. Louis sued for peace and was paid seven thousand pounds sterling to get out of England.

Until Henry III came into his majority in 1227, the government was administered by a group of experienced barons. Every now and then, especially when money was wanted for the royal treasury, the barons would have Henry reaffirm his commitment to the Charter of Liberties, and everybody got used to the idea that the barons had a say in making government policy.

No Englishman born into this period could miss the lesson of this outcome. It was possible to assail even so fundamental a principle as the divine right of kings and get away with it. With this barrier down, nothing seemed impossible.

England, separated from the mainland of Europe, provincial, uncultured, and dismissed by the more sophisticated French and Italians, could now question and test old ideas in splendid isolation. With Magna Carta upsetting the traditional political order, the country also became the perfect breeding ground for a new brand of learning, a new approach to the natural world that would spread and eventually threaten the social structure of Europe and the foundations of Christianity.

No one better embodied the character and spirit of this particularly English movement, or would surface more as both its champion and victim, than Roger Bacon. His parents had ambitions for him, an obviously gifted adolescent. In this new political climate, a talented boy, if educated properly, might advance beyond limitations of rank, perhaps even obtain employment with the king. So when he was fourteen, his parents sent him to school.

Had he lived even a half century earlier, education would have meant scriptural study at a cathedral school or monastery. But this was 1228, and a new institution had emerged, one that was exerting a great modernizing force over the whole of medieval society. Still in its infancy, it was already drawing away the best students and teachers, the most committed, the most intelligent, the most ambitious. So Roger Bacon's parents sent him off to attend one of these new institutions of learning, called a "university."

CHAPTER TWO

# Logic and Mysticism: Aristotle, Plato, and Christianity

• • •

THE UNIVERSITIES WERE THE STAGE on which the great approaching struggle between science and religious dogma was to be fought. That the universities had come about at just this time was due to a remarkable group of people that history has dubbed "The Translators," intermediaries between what was to become the most significant institution in the Christian world and the vast, luminescent Arab Empire.

In the century after the death of Mohammed in 630, Islam had grown with astounding speed, and by 750 the Baghdad Caliphate stretched from Persia and central Asia in the east, across North Africa to the Atlantic in the west, and up into Spain. Although titularly ruled by the caliph, this empire soon became a loosely united group of mini-empires of Turks, Persians, Berbers, Afghans, Syrians, and scores of others, each with its own language, customs, and ethnicity. However, just as Latin was the accepted language of scholarship of the north, Arabic became the accepted language of scholarship of the south.

As early as 830, Caliph al-Mamun established a center in Baghdad, called the "House of Wisdom," in which scholars were charged with translating into Arabic any works by Greek masters found within the vast boundaries of the caliphate. These included an array of texts by Hippocrates, Galen, Euclid, the Pythagoreans, Ptolemy, Seneca, Plato, and Aristotle. Sometimes these texts were in the original Greek, but often works were found already translated into vernacular languages, such as Syraic.

From this storehouse of classical knowledge the scholars of the Arab Empire went to work. Brilliant thinkers such as the mathematicians al-Kindi from Arabia and al-Khwarizimi, an Iraqi from whom we get the words *algebra* and *algorithm,* the metaphysicists al-Farabi, a Turk, and Avicebrol, a Jew, al-Hazan, a Persian, in optics, and Maimonedes, another Jew, in medicine kept creative thought and scientific inquiry alive for four centuries while Europe floundered in an intellectual morass.

In 1085, Christian forces under Alfonso VI retook Toledo for the west after almost four hundred years of Muslim rule and declared it the Spanish capital. Fortified, on a hilltop about seventy miles south of Madrid, Toledo, now better remembered as the setting for El Greco's cloudscapes than for its tradition of enlightenment, had been one of the intellectual centers of the Arab Empire. It was home not only to its Muslim population but also to one of the largest and most thriving Jewish communities in Europe. Demonstrating an insight that future Spanish rulers would lack, Alfonso and his heirs encouraged tolerance and coexistence among the religions.

In the early 1100s, Raymond, Toledo's archbishop, realized that there was treasure adrift in his city. His Arab predecessors had left piles of manuscripts behind both from the Greeks and their own scholars. Raymond, like al-Mamun before him, organized a school of translators, this time for the purpose of producing Latin versions of all the great works of science and philosophy that they could lay their hands on.

This was not a straightforward operation. It was laborious and involved a daunting polyglot of languages and dialects. While many of the Greek texts had been translated into Arabic, some were still in vernacular languages, some in Greek, and others in Hebrew. Even those already in Arabic presented a challenge for a Europe that had largely shunned all languages but its own.

Into this breach stepped the Jews, the most advanced linguists of the age. In one of history's most obvious examples of the advantages of tolerance over persecution, like "fertilizing subterranean streams," as the historian Will Durant put it, Jews shuttled between diverse societies, making available the information that would create a vibrant Christian scientific culture. Word of the effort got out, and European scholars flocked to Toledo, eager to get a first look at the material and add their expertise. As each new manuscript gained a Latin translation, scribes were employed to make copies and send them north.

So enormous was the volume and so disparate the subject matter that for the first time in Christian history, scholars were forced to specialize in one form of learning or another. There was a lot of overlap—geometry, for example, had a direct bearing on optics, which in turn had a direct bearing on astronomy. Specialists in one field had need of specialists in others. It was natural that scholars of different subjects, or even within the same subject, would begin to gravitate toward one another. By degrees a handful of cities attracted attention as centers of learning, which in turn drew more scholars and more manuscripts, and in each the embryo of a university was formed.

As a result of the new influx of information, for the first time in almost a millennium, scientific thought seemed poised to overpower the forces of ignorance and superstition that had dominated Europe through the Dark Ages. But would the Church allow scientific or intellectual inquiry to exist outside a strict, literal, and conservative interpretation of scripture?

Each side of this issue claimed to represent the true spirit of Christianity, and each used as its most pervasive and powerful weapon the words of a man, effectively an atheist, who had died more than three hundred years before Christ was born.

ARISTOTLE WAS THE LAST OF A STRING of three brilliant philosophers whose lives spanned what was certainly the most important hundred years in the history of human thought. Although the range of subjects that Aristotle studied and in which he came to be considered the authority was astounding—everything from botany to metaphysics, zoology to astronomy, poetry to politics to how to have better sex—his contribution was not to be found in any particular discovery. In fact, he often turned out to be wrong about specifics.

What Aristotle did achieve was the creation of logic itself, a method of ordered thought and analysis, of classifications and subclassifications, that remains the basis of most natural and social science to this day. Logic's essential building block, the syllogism (all Aristotelian conclusions are correct; that logic is a route to knowledge is an Aristotelian conclusion; therefore logic is a route to knowledge), is Aristotle's invention.

Aristotle's method of viewing the physical and metaphysical has survived intellectual challenge, mysticism, religious conflict, ludicrous parsing, and misapplication. His influence is still felt today by every student, businessperson, and politician—anyone who has ever written a book report or done an outline in bullet points or enjoyed Sherlock Holmes is walking in Aristotle's shadow. His contribution has become so basic, so fundamental to every aspect of scientific endeavor, that many of those reading him today might say, "So what? Everybody knows that."

The very quintessence of the Greek philosopher, Aristotle was not considered a true Greek at all, but rather a kind of provincial, having been born in Stagira in Greek Macedonia in 384 BC. His father, Nichomachus, was court physician to the king of Macedonia and had used his position to amass an impressive real estate portfolio. When Nichomachus died, Aristotle went off to live with his cousin and began to spend his sizable inheritance on decidedly unphilosophical pursuits. At seventeen, already bored with mindless pleasure-seeking, Aristotle journeyed to Athens—that was *really* Greece—and enrolled in Plato's Academy.

Plato, the scion of one of Athens's most distinguished political families, had turned to philosophy after unsuccessful flings at poetry and qualifying for the Olympics. (His real name was Aristocles, but he had acquired the moniker "Plato," which means "broad," as a result of his prowess as a wrestler.) While looking for something else to do with his life, he happened to meet the great Socrates.

(In fact, we can only assume that Socrates was great, since he confined himself to the spoken word and never wrote anything down. Most of what we know of his philosophy is from Plato's dialogues in which the character Socrates has any number of extremely profound things to say. The use of dialogue is in itself Socratic—Socrates taught by a question-and-answer method called the dialectic, in which an argument was proposed, then either supported or destroyed during a process of give-and-take.)

Plato studied with Socrates for the next nine years. After Socrates was forced to commit suicide by the supposedly democratic rulers of Athens, Plato wandered around a bit, was sold into slavery then ransomed by friends, and finally set up a school in the Grove of Academe, just outside Athens. Over the next forty years, he proceeded to produce some of the most provocative, profound, and beautifully written philosophy that the world has ever seen.

The core of Plato's philosophy revolved around a duality pitting necessarily flawed sensory perceptions against an unknowable reality that was composed of Ideals or Forms. A person looking at a green chair, for example, perceives only an imperfect conception of the idea of a chair and an equally imperfect conception of the idea of greenness. True green and true chairness are beyond his or her capacity to understand. Wisdom is defined merely as progress on this road to the Ideal, as well as the awareness that the Ideal exists in the first place. Plato had no real place for an interactive deity in his construct—God, as He would later be conceived of by Christians, would simply have been the sum of the Forms.

One surprising aspect of Plato's otherwise subjective construct was his reliance on mathematics, which in his day meant geometry. Over the gate of his school was written, "No one ignorant of geometry shall enter." To Plato, every Ideal, and thus all that made up the world, had a mathematical base and was held together by geometric elegance.

There were times when Plato's Ideals—geometrically elegant

or no—led him down some slippery paths. The work for which he is best known, the *Republic*, culminates in his call for rule by a philosopher-king. In the *Republic*, personal possessions are eliminated, there is no need for marriage among the elite, and children are taken from their parents at birth to be parented, in effect, by the state. Until age twenty, they would be educated in gymnastics and martial music. Those who showed promise would then be educated in astronomy and mathematics. (Those who did not were sent off to be menials.) After ten years, those who could not handle the schoolwork were sent to the military, and the survivors studied philosophy. At age thirty-five, the budding philosophers studied the practical aspects of government for the next fifteen years. Finally, when the generation reached age fifty the pyramid would be complete when the best of the group was chosen and the philosopher-king took his station to rule firmly but fairly, with wisdom, empathy, and justice.

The *Republic*, like most of Plato's work, is a dialogue starring Socrates, who here set down his vision of a just society, although whether or not Socrates ever advocated this odd mixture of fascism and communism is open to some question. That in practice this sort of autocracy was not the least bit Ideal but almost always degenerated to some form of ruthless dictatorship was not lost on future generations of scholars.

When Aristotle arrived, the Academy, as the school was called, was at the pinnacle of Greek intellectual life. It did not take long for Plato (and everyone else) to realize that his new enrollee was, to say the least, unusual. Aristotle quickly ceased being a student and became what amounted to a faculty member. It also did not take long before Aristotle found some significant areas of disagreement with his boss. Principally, he thought the notion of

Forms as the highest plane of knowledge was hooey. He did not think much of the philosopher-king idea either.

From there, in reasonably short order, Aristotle and Plato found that they disagreed on almost everything. Plato was fond of higher, ethereal, unknowable truths, and Aristotle thought knowledge began with and flowed from that which could be observed. Plato, the former wrestler, was a manly man who dressed simply. He probably growled. Aristotle, on the other hand, was skinny, dressed in the most fashionable togas, wore lots of rings, and spoke with a lisp. Plato is known to have referred to him as "a mind on legs," and scrawny legs at that. Aristotle, who had the money to do so, had also amassed a vast private library far grander than that of Plato, a disparity that was unlikely to draw the men closer. Reports that the two grew to loathe each other are probably overstated, but the notion of mere good-spirited intellectual rivalry is almost certainly too mild.

One thing that each of these philosophers never did, however, was underestimate the ability of the other and, when Plato died in 347 BC, empiricism or no, Aristotle assumed that the leadership of the Academy would fall to him. When he found out that Plato had instead granted the title to his own cousin, the foul-tempered Speusippus, whom Aristotle considered a dolt, Aristotle packed up and left. (Speusippus eventually justified Aristotle's faith in him by being forced to commit suicide after he was humiliated in a debate by Diogenes the Cynic.)

Aristotle headed back across the Aegean to where he had grown up and went to work for a minor king, the floridly named Hermias the Eunuch. Hermias, whose disability did little to improve his humor, was an authoritarian tyrant, but he also harbored ambitions to bring the flower of Greek culture to the déclassé provinces. He welcomed Aristotle, even encouraged him to marry his either niece, daughter, or former concubine, Pythias, who was twenty-one years younger than the great philosopher.

Aristotle moved around a bit after his marriage. Then Phillip of Macedonia, whose father had been treated by Aristotle's father, asked him to become tutor to his son, the thirteen-year-old Alexander, later to be known as "The Great." It must have been frustrating for one of the greatest minds in history to attempt to instruct a young, single-minded, bloodthirsty maniac in the niceties of higher thought. After four years, Aristotle gave up, returned to Athens, and, after getting passed over once again for the top post at the Academy, started his own school near the Temple of Apollo Lyceus (Apollo as a wolf), which he called the Lyceum (thus providing a convenient name for a theater in virtually every major city in the English-speaking world).

It was at the Lyceum that Aristotelian thought flowered. Aristotle taught in a method much like Socrates—he walked about, followed by his students, lecturing and engaging in discussions of whatever Aristotelian concept happened to be on the agenda. (The term *peripatetic,* meaning "walking about," came directly from the Lyceum.) Though Aristotle, unlike Socrates, wrote things down, almost nothing of what he wrote for posterity survives. Most of his sometimes cryptic, often ambiguous lecture notes, however, were saved.

Aristotle's concept of God, such as it was, was completely consistent with his methodology and, not surprisingly, rejected the vagueness of Plato's Ideals. Everything, he postulated, was based on motion, and nothing moves unless it is acted on by something else. But in order for this great progression to begin, there must be something that caused motion without having been acted upon itself. This Aristotle called the First Mover or Prime Mover. In the Aristotelian system, God (as the Christians later conceived Him) was reduced to whosoever had pushed the boulder downhill.

Although Aristotle's cold-logic empiricism was almost immediately adopted as the only way to do science—Galen, Ptolemy, and all the other great Greek and Roman scientists were Aristotelians—it occurred to some that Plato's concept of Ideals would be much better suited to mysteries of the infinite and the eternal. The duality of the knowable and unknowable (which would later take shape in Christianity as the conflict between reason and faith) allowed Christian, Jewish, and eventually Islamic theologians to allow for man's inability to perceive the essence of God. Aristotle, on the other hand, both literally and figuratively, lacked soul.

BY THE SECOND CENTURY A.D., Christianity had already been established as a dominant religion across much of Europe, Asia Minor, and northern Africa. Still, many Christians did not have a firm idea of the fundamental tenets of their religion. Even the concept of God itself seemed to have any number of different references in the Bible. Was God the Father in heaven, Christ himself, or some other, more ethereal overriding presence? There was also the question of what made Christianity different from other monotheistic religions. The words of the apostles provided direction but not logic, leaving its followers willing but uneasy. Christians needed their own thinkers to give the religion a philosophical base, but in their absence, they cast around among non-Christians to try and provide some answers.

Eventually an Egyptian named Plotinus, after studying ten years with an unemployed dockworker and lapsed Christian named Ammonius Sacchus, asked the question "How could such a flawed corporeal universe spring from the perfection of God?" This was the same question with which many Christians had been wrestling. To answer it, Plotinus extended Plato's duality, added God (sort of), and incorporated both into a movement that came to be called Neoplatonism.

God, Plotinus postulated, tap-dancing around the lack of a specific deity in Plato's work, was merely the Ideal, the unknowable, eternal One, The Good, from which all earthly things emanated, in the predicted imperfect form. From The One came The Intelligence, a kind of all-encompassing reality that in turn engendered The Soul. The Soul, unlike the first two layers, was active, and created all the lesser, individual souls that made up the earth.

To Neoplatonists, God was not an active being who created the universe in a voluntary act but rather merely a contemplative deity. As one moved down the ladder, reality became increasingly material, active, and imperfect. Nonetheless, each individual soul, flawed or besotted though it might be, was still part of The Soul, and therefore The Intelligence, and, ultimately, The One. All the different layers aside, Plotinus succeeded in transforming the concept of Plato's Ideal into a specific deity, which was exactly what Christians had been looking for.

Although his notion of a passive, contemplative God was branded with the epithet "pantheism," as were all theological constructs that either stated explicitly or implied that knowledge of God was present equally in all beings (a term the *Catholic Encyclopedia* today dismisses as "simply atheism"), Plotinus's concept of the existence of different forms was used by the Nicaean Council in 325 A.D. to create the Trinity, thereby reconciling the inconsistent definitions of God in the scriptures.

But Plotinus, Neoplatonism, and Plato gained their most significant partisan about fifty years later, when Christianity produced perhaps its greatest theorist ever, St. Augustine of Hippo. Through Augustine, who unlike Plotinus possessed impeccable credentials as a Christian, duality was fully incorporated into scriptural theory.

Augustine was one of a number of great figures in Christian history who began life as hell-raisers, but the others did not see fit to set down a record of their lascivious behavior for posterity. Born in 354, Augustine chronicled his early adult years of carousing in *Confessions*, which he wrote after his conversion. *Confessions* was ripe with phrases such as "hell's black river of lust," "filth of lewdness," and "putrid depravity," but despite the florid prose is lacking in graphic detail. At one point Augustine wrote the famous line "Lord, give me chastity—but not yet."

Augustine's transformation occurred in his twenties and was a great relief to his overbearing but fanatically pious mother, Monica. After gaining a reputation as a philosopher, he was invited to Rome. He turned down the honor in order to remain as bishop of Hippo (present-day Annaba, on the Mediterranean coast of Algeria, just west of the Tunisian border), a post he had been granted by the townspeople after the death of the previous bishop. He stayed at Hippo for the rest of his life. His last years were spent formulating defenses for the city, which was under threat from barbarian tribes that had already all but destroyed what was left of the Roman Empire.

St. Augustine was an unabashed Neoplatonist, lifting duality almost whole from Plotinus. In *City of God*, one of the most important Christian works after the Bible itself, Augustine created a clear separation between the perfection of heavenly things and the corruption of the earthly. In *De Trinitate,* Augustine once again used duality to establish the Trinity once and for all as a cornerstone of Christian faith. He used Platonic concepts to establish most of the other basic tenets of Christian theory that would survive until challenged almost one thousand years later by Thomas Aquinas.

But Platonic duality, while the perfect vehicle for Augustine, presented some unconquerable hurdles down the line. The problem with Plato, one that was to bedevil the Church throughout the Middle Ages and the Renaissance, was that his concept of Ideals—smoky, indistinct, and unknowable—could easily, perhaps inevitably, descend into mysticism, and mysticism has no inherent hierarchy. It doesn't take any special credentials to be a mystic, no particular prerequisite or method, no means of separating mystics who will see things your way from mystics who won't. Anyone, be it a pope, Francis of Assisi, Joan of Arc, Ignatius of Loyola, or the town shepherd, could claim with equal justification to have been spoken to by God. No training was required to receive a vision, and there was no means other than torture or repression to prevent a mystic from putting forth God's word however it was transmitted to him or her.

So, if dualism fell into the wrong hands, it could rob the Church of the role that it held most dear, and the one that formed the basis of its power—sole and unquestioned intermediary between man and God.

Plato, clearly, would not do.

What the Church required was a system within which it could claim objective authority in the interpretation of God's word and, by extension, the authority to prevent anyone with whom it disagreed from offering an alternative. What was needed here was science, logic, learning—all those specialties over which the Church could claim a virtual monopoly, separating it from the common man—or even from kings.

What they needed, in other words, was Aristotle.

CHAPTER THREE

# Logic and Theology: The Evolution of Scholasticism

· · ·

IN AUGUSTINE'S TIME, none of Aristotle's work was known to Christian scholars. In fact, out of the entire incredible range of Aristotelian writings, only one set of manuscripts had been recovered, and even that had not been translated into Latin. Fortunately for the future of Christianity, that set happened to contain writing on the one subject that the religion needed most—Aristotle's work on logic, the *Organon*.

From the *Organon* would come scholasticism, a system of analysis and teaching by which Christianity would leap forward and then be held back, at once the most progressive and reactionary innovation in Christian education and philosophy in its history. Scholasticism was the method that would be used in every university, the rule book in the battle of dogma against science, the system by which Roger Bacon learned and later taught, which molded his philosophy of science and provided its greatest impediment.

The process toward scholasticism began soon after Augustine's death in 430, when a Roman consul named Anicius Manlius

Severnius Boethius finally translated the *Organon* into Latin. Boethius—whether he was actually a Christian himself is unclear—was one of the top advisors to King Theodoric, but ruler of Rome was not the elite position it had once been. First of all, Rome itself was no longer the center of the empire. That was in Constantinople, where the Byzantine emperor Justin ran the show. Second, Theodoric was not a Roman but rather an Ostrogoth, one of the many tribes to the north that had once been colonized by Rome but, during the previous century and a half, had virtually annexed what was left of the empire. Even so, Theodoric, who probably never learned to read or write, saw himself as Roman and tried to carry on the great intellectual traditions of past emperors, such as Marcus Aurelius.

Boethius was an exceptional scholar who, because of his learning and Theodoric's cultural ambitions, became a favorite of the king, much as Aristotle had become a favorite of Hermias the Eunuch. Given a free hand, Boethius produced original works on mathematics and science, as well as his translation of Aristotle.

He did not have a chance to do much with it, however, since, as it turned out, he had become a bit too much of the king's favorite. Some of those at court who aspired to be favorites themselves convinced Theodoric that Boethius was part of a plot to kill him. Their proof was a document supposedly signed by Boethius, which Boethius claimed was a forgery. Theodoric, in no position to determine the genuineness of any document for himself, let his Gothic blood win out over his Roman veneer. He chucked Boethius into prison without trial. Boethius slowly rotted in a cell, writing philosophy and commentary. Finally, in 525, Theodoric had Boethius executed in a particularly unpleasant way. A cord was tightened around the condemned man's neck until his eyes popped out, and then he was slowly beaten to death. (Theodoric died two years later. He is said to have cried over the injustice he had perpetrated on an innocent man.)

Christian education, as it was to persist through the Renaissance, began to take form soon afterward. Flavius Magnus Aurelius Cassiodorus, a friend and pupil of Boethius's, read the Latin translation of the *Organon*, then produced a text, *The Course of Religious and Secular Studies*. In it, Cassiodorus folded Aristotelian logic into the old Greek method of dividing a curriculum into the seven liberal arts, grouped into a language *trivium* (grammar, logic, and rhetoric) and science *quadrivium* (arithmetic, geometry, astronomy, and music) and asserted that this was the proper way to teach Christians. Cassiodorus did not have a great deal of competition in the philosophy of education arena, and so the seeds of scholasticism were sown.

In 529, the new Byzantine emperor, Justinian, Justin's adopted son and an avowed Christian, closed the Academy in Athens, by then the last stronghold of Neoplatonism. That date is often considered the beginning of what has been called, with some justification, the Dark Ages. Still, although there was a virtual halt in creative thought in Europe, Boethius and Cassiodorus had nonetheless created an Aristotelian framework within which Christianity could function, as it did, almost unchanged for half a millennium.

During this time, scholasticism ever so slowly ripened—it took three hundred years just to get past the rote-based master-lectures-student stage. Instruction took place largely in monasteries and cathedral schools, and so one-sided was the process that it was the teacher rather than the method that was referred to as "scholastic."

Finally, in the tenth century, scholasticism began to acquire an intellectual pulse. Masters reached more deeply into the works of Boethius and Cassiodorus and reintroduced the Socratic dialectic as a teaching vehicle. When students became involved, the masters actually had to find answers to difficult questions every once in a while, and new and better arguments were forced into the process. This in turn led to more sophisticated questioning, which required more thought and study in the replies.

Within a century, scholasticism had matured into the most powerful tool for maintaining and perpetuating doctrine that the Church had ever seen. Almost in gratitude, Aristotle and his logic were adopted by Church fathers and the man himself (now called simply "the Philosopher") was said to be infallible. Even Augustine's (and Plato's) duality was now formalized and tempered on the Aristotelian anvil.

Regardless of the degree of sophistication, however, the basic aim of the scholastics had not changed. They remained uninterested in uncovering new knowledge, only in cementing the unlikely but now solid bond between Aristotle's logic and the Bible's revelation. Scholastics argued that since God was the supreme power in the universe, and that revealed truth (as set down in scripture) was His message to man, human reason was subordinate and, if ever it seemed to contradict revelation, must give way. In other words, if science and faith butted heads, either the science was wrong (most likely) or someone had made an incorrect scriptural interpretation. Scholastics came to call philosophy the servant of theology, because they used philosophy to understand and explain revelation.

In 1150, a teacher at the cathedral school in Paris named Peter Lombard institutionalized the entire system in a work titled *Four*

*Books of Sentences*. "Sentences" were the method pioneered by Cassiodorus to subject religious truths to the full dialectic treatment. Lombard now used this technique as the structure of his work. Taking what were then the standard divisions, he devoted the first book to God and the Trinity, the second to creation, the third to incarnation and redemption, and the fourth to the sacraments. In each book, he presented a series of propositions (*quaestio*), and then subjected each to dialectic examination (*pro* and *contra*). This process came to be known as disputation (*disputatio*). Neither Lombard's propositions nor his arguments were original—almost everything he cited came from St. Augustine—although he threw some more contemporary Church masters into the mix for topicality.

What Lombard achieved was the ultimate ecclesiastic textbook, a systematic, ordered, encyclopedic compilation of every important piece of Church dogma, complete with every objection that had been raised against it, to which Lombard supplied the appropriate counterargument. When there were counters to the counter, Lombard showed students how to answer those as well, and on and on until every proposition had been irrefutably cemented.

*Four Books of Sentences* became the standard Church text for the next two centuries, and soon after it was published it became impossible to gain a degree in theology in any cathedral school in Europe—or later, any university—without writing a commentary on the text. So it developed that theology itself became the ultimate science and the logic of Aristotle was employed to prove the very arguments the man himself had spent his life trying to disprove.

But just when it seemed that the Church could sit back and relax, secure in having tamed Plato, duality, pantheism, and other heterodoxy with Aristotelian logic, a body of new knowledge prepared to flood through Europe. Most of this knowledge came from a single source, one man, one philosopher whose works were so obviously true that it was going to take everything the Church had to hold him off.

It was Aristotle.

This time, he came from the Arabs.

WHILE EUROPE HAD ACCESS only to Aristotle's works on logic, scholars from countries across the Arab Empire possessed Arabic translations of the full range of the Philosopher's work. With the fall of Toledo and the subsequent efforts of the Translators, many of the lost writings of Aristotle came north and seeded the universities. Much of this material was scientific and therefore welcome as part of the new learning. But then there were metaphysical works, such as *De Anima (On the Soul)*. This segment of the Aristotelian corpus was less appealing to the Church, as it contained arguments on questions of man, God, and eternity that could not be blended quite so easily into Christian dogma as had the *Organon*. In fact, there were some specific passages that might even be seen as contradicting axioms of the Christian faith.

If that were not enough, two of the greatest Arab thinkers had done substantial analysis of Aristotle's metaphysical works—very persuasive analysis, as it turned out—and Latin translations of their commentaries came north as well. By the time Roger Bacon was sent off for his schooling, these commentaries had become as much a part of university learning—although often covertly—as the *Sentences*. While sometimes the work of these men lent support to Christian theology, more often it lent itself to reexamination of heretofore unquestioned precepts.

The first of the two was Abu Ali al-Hussain Ibn Abdallah Ibn Sina, who is generally (and more economically) referred to in the West by his Latin name, Avicenna. Avicenna was one of the great theorists in the history of Islam, and probably one of the greatest minds ever. He was born in 980 in Bukhara, in present-day Uzbekistan, then part of the decaying Samarind Empire. By age ten, largely self-taught, Avicenna had memorized the Qu'ran, and by thirteen was studying mathematics, medicine, and philosophy. By sixteen, he was treating patients, and one year later he had the good fortune of curing the Samarind king Nu ibn Mansur of an illness that all the other doctors had assured him was fatal. Asked to name his reward, Avicenna desired only to be allowed access to the Royal Library.

From there, he produced a large volume of commentary on Aristotle that philosophically was pure Neoplatonism. He ascribed to a hierarchal order of God and then a single active intelligence from which individual human personalities sprang. Instead of the vague concept of the One, however, Avicenna plugged in Aristotle's Prime Mover. For Christians, this would prove helpful because it established a method of retaining the unknowable ideal without sacrificing the scriptural truth that God created the universe consciously by an act of will. Still, if theologians accepted Avicenna's construct, it seemed certain that the old problem with mysticism was going to resurface. Moreover, Avicenna had stuffed more intermediaries between God and man than Plotinus had. As such, his theology never really caught on in Europe (unlike his *Canon of Medicine*, which became the standard medical text for the next five hundred years). Still, his Aristotelian contribution seemed to have helped solve a fundamental problem for Church scholars.

Helping along Christian theology, of course, was the farthest thing from Avicenna's mind. He was interested in a better Islam. His idea of attaining that end by using Greek thought rather than the Qu'ran itself not surprisingly made him any number of enemies. In 1150, all of his philosophical works were ordered burned by Caliph Mustanjid. (It is civilization's luck that by then the work of Avicenna had spread beyond the caliph's reach across the known world.) He was imprisoned twice (once escaping disguised as a mystic holy man), forced to flee more often than that, and died mysteriously in 1037, quite possibly after being poisoned.

While Avicenna, brilliant as he was, produced some Aristotelian commentary that made its way into theological debate, no Muslim in history had more impact on Christianity and Western civilization than the second great Arab philosopher, Abu al-Walid Muhammad Ibn Ahmad Ibn Rushd, who came to be known in Europe as Averroës.

Like Avicenna, Averroës was a physician. He was also a judge, astronomer, and counselor to kings. He was born into wealth and position in 1126, the son and grandson of chief justices in Cordoba. By that time, the cultural center of the Arab Empire had moved to Spain. As a young man already with a reputation as a prodigious scholar, he was brought before the emir in 1153 and asked if he thought the heavens were eternal. Averroës hesitated and then said he did not know.

He would spend the rest of his life trying to find out.

For more than thirty years, Averroës went through Aristotle line by line, translating, interpreting, and producing commentary.

Because only Aristotle's lecture notes survived, just what the Philosopher meant by this passage or that had often been open to question. Averroës's scholarship was so great, his knowledge of Aristotle so complete, that he became known simply as "The Commentator." In the slack moments when he wasn't working on Aristotle, Averroës had time to serve as chief justice in Seville, then in Cordoba, be appointed as personal physician to the emir, and discover both the function of the retina and that immunity to smallpox is conferred on those who survive the disease.

What Averroës concluded after his study was to shake the foundations of not only his own religion but Christianity as well. Like Aristotle, he believed that reason, not revelation, represented the highest plane of wisdom, a point of view with which Europeans had yet to contend, since they had until recently been restricted to the *Organon*. Averroës extended Aristotle's reason-over-revelation contention with the notion that philosophy—the science of reason—(and therefore wisdom) is only for the elite. For Averroës, the vast majority of humanity was by nature unequipped to master the subtleties of higher thought, and thus incapable of enlightenment. These people were, in fact, happiest and most fulfilled when told what to do and what to think. For them, faith and revelation would more than suffice, and on no account should philosophy or dialectic be introduced to confuse them. A second, far smaller group needed a more rigorous explanation of why one proposition was true and another was false. This group could be taught philosophy, but only in a more or less rote manner. Finally, there were those few actually capable of reaching a higher plane. For these special men only did reason provide greater insight than faith, and therefore only they should be permitted to study or teach philosophy. (If Averroës, the consummate Aristotelian, was aware of how close in this construct he was drifting toward Plato's *Republic,* he never let on.)

Although Averroës was a devout Muslim, he was not himself a theologian or even a philosopher per se. That is perhaps why, like Avicenna, although his aim was to integrate Aristotle and Plato into Islam, he looked more to the Greeks than to the Qu'ran for the basis of his work. Also as with Avicenna, this approach did not endear him to what was becoming an increasingly fundamentalist Islamic clergy. Averroës, enjoying the protection of kings, continued to work until, in 1194, Emir al-Mansur ordered all of his writings (except a small number that were considered pure science) burned. The emir also decreed that no one at all should read philosophy and that his subjects should throw all philosophical works into the fire. Averroës was sent into exile for heresy. He was recalled four years later but died soon afterward, in 1198, just as Innocent III was ascending to the papacy.

The death of Averroës coincided with a pivotal juncture in European history. For his own people, it marked the final downturn of empire. As the Arabs retreated further and further into religious fanaticism and intolerance, infidels began to carve up their possessions. Within decades, all Spain except Granada was lost to the Christians, crusaders took Jerusalem, and the Mongols under the Khans swept through Muslim Asia and attacked and destroyed Baghdad.

But as the legacy of Averroës faded along with the glory of the Islamic south, it was accelerating a new blaze of science and empirical thought in the Christian north. European scholars traveled to areas of Arab influence, control, or legacy expressly to seek new learning and then returned to the universities to pass it along. As a result, from an odd swirl of languages, religions, and nationalities, Europe became the beneficiary of the single most prolific infusion of knowledge in its history.

ALTHOUGH TOLEDO WAS THE FIRST SOURCE of the new learning, another kingdom to the west soon rivaled Spain for the importance of the translated material that it sent to the universities of Italy and northern Europe. It was the Sicily of the Holy Roman Emperor, Frederick II.

The Holy Roman Empire was a loose, often shifting confederation of kingdoms in central and southern Europe. Shaped by wars, alliances, and treaties, by the thirteenth century the Holy Roman Empire—described by Voltaire as "neither holy, Roman, nor an empire"—was primarily German, but it included the wealthy and strategically placed island of Sicily.

Entitled by birth to rule both Sicily and Germany, Frederick, grandson of the ferocious Barbarossa ("Redbeard") and heir to the Hohenstaufen dynasty, had been made a ward of Innocent III after the death of his parents when he was four. Innocent, seizing the opportunity to retain the empire for the papacy, shunted him off with guardians.

When Frederick was about seven, he began to be seen as a rallying point in Sicily by those opposed to papal rule. Some of those appointed by Innocent to "care" for the boy then attempted to kill him, but Frederick was snatched away by supporters. He was taken in and fed by one family after another and proceeded to grow up on the streets of Sicily, hiding from his powerful enemies and learning to live by his wits. Sicily was then the center of the world, the meeting place of all cultures—Christians, Muslims, Byzantines, Africans, and Jews—a wealthy, exotic milieu of tastes and smells, customs and tongues. Frederick, who had been taught to read somewhere along the way, acquired an affinity for history and mathematics. He learned Arabic and Hebrew, as well as Latin. By the time he was in his mid-teens, Frederick was already every bit the pope's match for intelligence, cunning, and ambition.

Then, when Frederick was eighteen, as a buffer against the ambitions of a would-be emperor, a hulking dimwit named Otto the Welf, Innocent, convinced of his own infallibility, granted the boy his birthright and declared him Holy Roman Emperor and ruler of Sicily.

Underestimating this blond, handsome teenager was to be the greatest blunder of Innocent's career. Frederick would grow into a scholar, warrior, poet, freethinker, mathematician, and scientist, and nearly achieve his grandfather's ambition to conquer Rome itself. He would come to personify secular power and thought, questioning the fundamental tenets of scripture and the rules by which medieval Europe lived. More important, Frederick, later referred to as "the Antichrist" by future popes, would provide a focal point for the scientific revolution in which Roger Bacon participated so significantly and come within a whisper of creating sufficient momentum for it to overpower a thousand years of Christian dogma.

The new emperor Frederick II quickly threw off the papal yoke and used Sicily's fabulous wealth to create the most flamboyant court in all Europe. He himself became known as *Stupor Mundi,* the wonder of the world. He traveled with a menagerie of exotic animals and birds, including leopards, panthers, bears, peacocks, doves, ostriches, a giraffe on a chain, and an elephant with a small tower on its back. He had slaves in silk, wagons of treasure, a variety of bric-a-brac rendered in gold and precious jewels. He had a harem and an army of Saracen slaves. For dinner he ate figs, dates, and nuts, all unknown to the northern palate. Northern European visitors were stunned by the opulence and Byzantine grandeur of his court. He might have been the emperor of Christendom, but he looked like the Sultan of Baghdad.

This appearance was no coincidence, because in his heart he was much more Arab than Hohenstaufen. Perhaps because he was the only Christian ruler with a firsthand knowledge of the Arab Empire, the only one to have actually met one of his Arabian counterparts, the only one to exchange letters and gifts, he was also the only one to really appreciate how advanced Arab civilization had become—to dare to think it more advanced than Christendom.

Frederick was personally the most educated monarch in Europe. He was conversant in at least seven languages and could read three or four. He devoured classics that had been translated in the previous century, and his knowledge of mathematics was superior to that of almost any scholar in Europe. He sent such complicated geometry problems to the sultan of Damascus that the sultan was forced to pass them on to his most advanced Egyptian mathematicians for solution. Another sultan, noting the emperor's interest in science, sent Frederick the gift of an astrolabe, used for measuring the altitude of the sun and the stars. Frederick was the only ruler in Europe to write his own book, called *On the Art of Hunting with Birds*. In it, he carefully detailed precise observations of different species and their habits, including migratory patterns. He once had a vulture's eyes sewn shut to test whether the birds hunted by sight or by smell.

He was so consumed by the new learning that in 1224 he established his own university at Naples on the Italian mainland and actively recruited scholars, poets, painters, and scientific thinkers to his court for the purpose of translating and studying scientific works. One of these, an English astrologer, would be as responsible as anyone in history for bringing the blessing and curse of Aristotle to Europe.

Michael Scot had been a translator in Toledo before relocating to Frederick's court. He was equal parts scientist, philosopher, and quack. Frederick was so impressed with Michael's knowledge of Arab scholarship that he appointed him court astrologer and general all-around sage. With the emperor's encouragement, Michael Scot practiced alchemy and conducted experiments, detailing his observations in a scientific manner.*

Michael Scot's most important contribution, however, was his translation into Latin from Arabic of Aristotle's works on natural science, the *libri naturales,* such treatises as *On the Parts of Animals, On the Generation of Animals,* and the *Physics*. He also translated more provocative texts—the *Metaphysics, On the Heavens,* and *Ethics*. More than that, he included his own translations of the commentaries of Avicenna and Averroës.

These were just the sort of secular subjects that Frederick held dear and Rome feared. Michael's translations made their way to Paris, and would be the very translations later used by Albertus Magnus and Bacon himself to propound their views on experimental science. "Although only some of his works on logic and certain others have been translated from Greek by Boethius, yet

* Alchemy, although now discredited as fakery performed by would-be sorcerers wearing cone hats embellished with stars and crescents while trying to change base metals into gold, was actually a legitimate and clever means of explaining the natural world. Alchemists simply stated that all matter was made up of like material in varying densities and proportions, and it was therefore possible to change one form of matter into another. Modern scientists do just that with atomic fission.

from the time of Michael Scotus, whose translations with authentic expositions of certain parts of Aristotle's works on nature and metaphysics appeared in the year of our Lord 1230, the philosophy of Aristotle has grown in importance among the Latins," Roger Bacon was to write later.

The combination of Michael Scot's translations and the commentaries of Averroës shook Christian orthodoxy. There were indeed some passages in this new Aristotle that called into question not only the old, accepted, logical Aristotle but also some fundamental tenets of the Christian faith. Three elements in particular seemed devastating.

First, Aristotle said quite specifically in the *Metaphysics* that the universe was eternal, with no beginning or end, that it had simply "always been." The Bible, on the other hand, said equally specifically that God had created the heavens and the earth in a finite period by an act of will. Second, Aristotle claimed that after death an individual's soul blended into what he called "the tenth intelligence," a kind of generalized soul that was surprisingly reminiscent of the discredited Plotinus. This was taken (correctly) to mean that Aristotle denied the immortality of the individual soul as Christians recognized it, which therefore rendered the concepts of heaven and hell meaningless. Without the promise of heaven—or the fear of hell—the authority of the Church to create and enforce standards of behavior would erode considerably.

The big one, however, the problem that seemed utterly irreconcilable with the Christian faith, was the Aristotelian division (as interpreted by Averroës) of truth into two distinct forms—that which could be known by reason (philosophy) and that which must be accepted by faith (religion). While the division of reason and faith was not in itself repugnant to Christians—it had, after all, been a cornerstone of St. Augustine's work—faith being *subordinate* to reason, consigned as the *lesser* truth (as Aristotle seemed to have clearly stated), was anathema. If philosophy was to represent the highest level of knowledge, what would happen to the rule of God and, worse, the rule of God's agents on earth—the pope and his cardinals, bishops, and priests?

In short, what Aristotle presented in these new translated works, in vastly greater detail and breadth than the old, was a philosophy of science itself, a theory of how to make sense of it all and use the knowledge for human advancement. The man who had given the Church not only its method of learning but also its fundamental truths of nature, such as the geocentric universe and the composition of matter as mixtures of the four basic elements of air, fire, water, and earth, was now saying that the fundamental tenets of Christianity—to say nothing of the Bible—that the very manner in which Christians viewed this world, and the world beyond, were *wrong*.

Michael Scot's translations reinforced the already powerful pull toward secular knowledge that had begun with the translated manuscripts from Spain. The new knowledge from the south was proving both irresistible and inexorable in the north, and for the first time large numbers of students from across Europe came to cities such as Paris to study Aristotle rather than the *Sentences*. That, in turn, led to a radical increase in the number—and influence—of masters who taught only science or the arts and often were not even members of the clergy themselves. Still, control of education had always been a cornerstone of ecclesiastic power, and, since in the past no one outside the clergy had been much interested in schooling, many theologians felt that it would be a simple matter to seize control of the universities.

CHAPTER FOUR

# Dogma, Drink, and Dissent: The University of Paris

• • •

BY THE TIME ROGER BACON WENT TO SCHOOL IN 1228, about ten universities had been founded across Europe. Of these, the undeniable apex was Paris. Paris had been the beneficiary of a happy combination of convenient geography, a pleasant climate, status as a capital city, and an enthusiastic, devout line of kings. "The Italians have the Papacy, the Germans have the Empire, and the French have Learning" went a medieval saying.

For all this Learning, however, it was necessary to have Students, and students were (and still are) a mixed blessing. By the 1190s, students and masters composed more than a tenth of Paris's total population of about 30,000. As education turned secular, more and more students entered school specifically to gain an undergraduate arts degree and become doctors, lawyers, or clerks. They came from all over Europe, and from every stratum of society. There were noblemen and peasants, French, Germans, Italians, Spanish, English, and everyone in between. The vast majority of this new transient population was between the ages of fourteen and eighteen—and they were all boys.

Paris very quickly developed into a classic college town. Twenty-first-century parents will be interested to learn that university life has not changed very much in nearly a millennium. The overwhelming preponderance of letters home, for example, were pleas for money. Since many of the fourteen-year-olds who arrived at the school had not yet had the chance to learn to write, there were the equivalent of form letters for the purpose of conniving money out of parents or patrons, with blank spaces for the student's name and his target. "A much copied exercise contained twenty-two different methods of approaching an archdeacon on this ever delicate subject," observed Charles Haskins in his meticulously researched *Studies in Medieval Culture.*

A sample letter, composed by a teacher for the benefit of his students (upon whom he was reliant for his fees), went as follows:

> *I know not what to offer you, my sweet father, since I am your son, and after God, entirely your own creature—so completely yours that I can give you nothing. But if I can remember what the child's instinct prompts it to say, I might sing, as the cuckoo incessantly sings, "Da, da, da, da": and this little song I am compelled to sing at this moment, for the money which you gave me so liberally for my studies last time is now all spent, and I am in debt to the tune of more than five shillings . . .*

Hormones were also a major problem. There was quite a lot of drinking and brawling. The undergraduates fought with themselves, with the townspeople, with the masters, and with their servants. The students were *so* unruly, in fact, that sometimes the masters were forced to close down the university altogether. There was a great deal of interest in women as well. Since many of the city's landlords refused to rent to students or provide teaching space, classes were held in the seamier sections of town, often over brothels. This made the Paris prostitutes a kind of adjunct faculty.

A famous thirteenth-century minstrel named Rutebeuf summed up the situation this way:

THE SONG OF THE UNIVERSITY OF PARIS

*Much argument is heard of late,*
*The subject I'll attempt to state,*
*The student-folk of Paris town*
*(I speak of those in cap and gown,*
*Students of art, philosophy—*
*In short, "the University,"*
*And not our old-time learned men)*
*Have stirred up trouble here again.*

*To give his son a chance to stay*
*In Paris, growing wise each day,*
*Is some old peasant's one ambition.*
*To pay his bills and his tuition*
*The poor hard-working father slaves;*
*Sends him each farthing that he saves,*
*While he in misery will stay*
*On his scant plot of land to pray*
*That his hard toil may help to raise*
*His son to honor and to praise.*

*But once the son is safe in town*
*The story then reads upside down.*
*Forgetting all his pledges now,*
*The earnings of his father's plow*
*He spends for weapons, not for books.*
*Dawdling through city streets, he looks*
*To find some pretty, loitering wench,*
*Or idle brawl by tavern bench;*
*Wanders at will and prie about*
*Till money fails and gown wears out.—*
*Then he starts fresh on the old round;*
*Why sow good seed on barren ground?*
*But swaggering hauberks, as they sit*
*Drowning in drink their feeble wit;*
*While three or four of them excite*
*Four hundred students to a fight,*
*And close the University.*
*(Not such a great calamity!)*

*Why send a boy away to school*
*There to become an arrant fool?*
*When he should be acquiring sense,*
*He wastes his time and all his pence,*
*And to his friends brings only shame,*
*While they suppose him winning fame.**

The animosity between the university and the rest of the city grew steadily until, in 1200, the school won a decisive victory over

* Translated by Marion E. Markley from the French poem of Rutebeuf, written in octosyllabic couplets, about the middle of the thirteenth century.

the citizens of Paris. The incident began when a German student's servant went down to the local pub to purchase some wine for his employer. There was a disagreement over price, and the tavern owner insulted the servant. The servant went back to the German student empty-handed. The student rounded up some of his compatriots and took them to the bar, where they proceeded to trash the establishment and beat up the owner. There was outrage in the city, and the civilian authority, under the command of the royal provost, got up its own mob, went over to the German student's quarters, and in retaliation killed a bunch of university people, including the German student.

The masters, who understood their growing power in a city that was becoming more and more dependent on the university as an industry, called on the king to act against his own provost. They threatened to close the school and initiate a mass exodus of the faculty if their demands were not met. Philip Augustus was in power, and he ruled without hesitation in favor of the masters. Not only did he throw all of the Parisians involved in the incident, including his own provost, into prison for life, but he seized their lands, burned their houses, and then pursued those who fled from the city across France and brought them back for the same punishment. Just to make sure nobody crossed this line again, Philip Augustus issued a charter giving all scholars clerical status, which meant that they could not be tried in civil courts, nor could their property or persons be seized.

In the end, though, the primacy of the University of Paris was secured not by the kings of France but by its most powerful alumnus, Innocent III. Innocent, who had seen the potential of the University firsthand, moved quickly to make Paris the vehicle with which to provide Rome with a steady stream of superior theology graduates. In 1209, he issued a charter confirming what amounted to a guild of masters and essentially placed it under papal protection, a policy that would be continued by subsequent popes. From that time on, the University of Paris became the papacy's school, the acme of theological study, and, most important, a prerequisite to high ecclesiastic office.

Innocent and the theological faculty were in fundamental agreement over what should and should not be taught, but the arts faculty had different ideas. All those brawling, drunken students, who came to the school not to study theology but for degrees in law or medicine gave the arts masters significant clout and helped them grow into a political force. By sheer power of numbers, the arts faculty could now battle on equal terms with the more conservative theology masters. At the time, no one anticipated that this battle would evolve into the philosophical struggle for the soul of Christianity that it was to become, with the arts masters—of whom Roger Bacon would become the most prominent example—championing science and a more intellectually inclusive Church and the theology masters trying to hold back the dual tides of reform and secular knowledge.

The rift began in earnest in 1210, when Peter of Corbeil, the archbishop of Sens, who had been Innocent's own theological master at Paris two decades earlier, banned a number of books representing the new learning, among them Aristotle's works on natural philosophy (as science was known through the Middle Ages and into the Renaissance) and Avicenna's commentaries. The banned books were not to be "read at Paris in public or secret." This ban would be reissued periodically, but to no avail. The arts faculty continued to read and debate Aristotle and his commentators. It seemed that even Innocent's great personal power would not be enough to prevent the arts masters from teaching whatever they liked, and thereby undermining his plans for the future of the Church.

Then, suddenly, fate threw a wild card into the mix. Two new religious orders came into being. Each embraced as its primary goals poverty, charity, and a simple Christlike existence, yet each would nonetheless recruit among its members the best scientific minds of the time and turn them loose on the universities. One of these orders would come to be known for exalting Albertus Magnus and Thomas Aquinas, the other for suppressing Roger Bacon. Each order was referred to by the name of its founder—St. Dominic de Guzman and St. Francis of Assisi. The Dominicans and the Franciscans would entirely alter the balance of power both in the universities and in the Church at large and determine the course of science for the next four centuries.

ST. FRANCIS OF ASSISI IS ONE OF THE BEST KNOWN, best loved, and most influential figures in all of religious history. Born in Italy in 1182, the ne'er-do-well son of a wealthy businessman, as a child he dreamed of the romance of chivalry and knighthood. He was given a cursory education in Latin and set up in his father's business, to which he paid little attention, preferring to spend money rather than to earn it. His life was gay, generous, and entirely frivolous.

His conversion came swiftly and irresistibly. The emptiness of his existence struck him, and about 1206, when he was in his twenties, he began to act strangely. He for whom "the sight of lepers was so bitter in the days of vanity that he looked at their houses two miles off and held his nose," astonished his friends by kissing one; soon after he stole money from his father in order to give it to an impoverished priest. For this act he was renounced by his family and, left entirely to his own devices, began to beg for lepers, the poor, and himself.

In 1209, while listening to the Gospel being read at church, he had an epiphany. As the priest read from Matthew, "As ye go preach, saying, the kingdom of heaven is at hand . . . Possess neither gold nor silver nor money in your purses, no wallet for your journey, nor two coats nor shoe," Francis cried out, "This is what I am seeking!" From that time on he devoted himself to poverty and preaching, seeking to emulate the simple life of Christ. The force of his faith, in combination with his appealing personality and gentleness, brought him a motley crew of twelve followers, including a nobleman, a peasant, and an idiot, and, being informed that he needed the pope's approval to start an order, Francis journeyed to Rome to see Innocent III.

At the time, heretical sects were on the rise; a German bishop estimated that there were 150 operating in Europe. This increase was a threat not simply to the faith; the sects were seizing Church property and controlling sections of Europe as if the Church did not exist. Of these, the one that represented by far the greatest threat was in the south of France. Known as Cathars or Albigensians, like Francis of Assisi they preached the simple life (although they did not believe in begging), a direct response to the excesses of a Church that allowed bishops and legates to live like kings. The Albigensians did not recognize the authority of the pope, and developed their own rituals.

Realizing that the Cathars drew their appeal from the revulsion of common people to the excesses of Church officials, Innocent had already sent a Spanish bishop, Diego de Acevedo, and his canon, Domingo de Guzmán, to Languedoc, the French province that was the Cathars' spiritual center. The two Spaniards, both with a reputation for simple piety, imitated the Cathar "perfects" (priests) and traveled about like apostles, on foot, going door to door, begging bread and preaching the Gospel. Soon the bishop died, and Domingo, or Dominic, as he was known in Latin, continued alone, amassing a group of followers who now had a pious alternative to the Cathar heresy.

But one man could do only so much, and the number of Cathars swelled. At this point, Innocent did something unprecedented in the annals of Christian history: he called for a crusade, not against the infidel Muslims for the recapture of Jerusalem, but against fellow Europeans. Encouraged first by Philip Augustus and then by his son, Louis VIII, who saw this as an opportunity to swallow up huge sections of the south, the crusade was conducted with as great a ferocity as has ever been perpetrated in the name of God. Year after year, knights and soldiers from the north, led by Arnold Amaury, a Cistercian monk, and the ruthless warrior Simon de Montfort, the second son of an impecunious but famous noble French family, swept down the countryside, surrounding and besieging Albigensian men, women, and children. If the Cathars surrendered, they were slaughtered or horribly mutilated.

In the face of all this, it took a pope with the clear-sighted pragmatism of Innocent to recognize that Francis of Assisi's request for a new order based on total renunciation of the temptations of worldly life was not another heretical threat but an opportunity. Francis was committed to unconditional, absolute obedience to the pope, so Innocent gave him the authority to preach and establish the Order of the Friars Minor (Little Brothers).

It is astonishing how quickly the ideas of the Franciscans caught on. Everywhere, noblemen gave up their titles to wash the feet of lepers; businessmen surrendered their wealth to beg for themselves and the poor. Lawyers and peasants alike donned the coarse brown or gray robes tied with cord around the waist that were the trademark raiment of the mendicant order. Every friar, from the lowliest novice to the minister general, eschewed any form of medieval transportation and walked, barefoot, no matter what the weather. It is as strong a testament to the human desire for faith as has ever been recorded.

Dominic, too, had applied to Innocent for permission to establish a new order. Dominic did not actually convert many Cathars—the brutality of the crusade saw to that—but still, he received his authority in 1215. Like the Friars Minor, Dominic's group, the Order of Friars Preachers, were mendicants whose assignment was to travel the world preaching the word of God for the salvation of souls. They too adopted a uniform—a white robe covered by a black cloak—and walked everywhere.

Francis began with just twelve acolytes; Dominic had sixteen. Both sent their followers out into the world to carry out their missions. It was not always easy for the friars. Not everyone embraced the mendicant ethic. There were difficulties also with foreign languages and customs, as Brother Jordan, an early Franciscan convert, discovered:

> *To Germany there were sent . . . John of Penna with about sixty or more Brethren. When they entered the borders of Germany, and, not knowing the language, were asked if they wanted shelter, or food, or other things of the sort, they replied "Ja," and in this way they received a good welcome from divers folk. When they saw that by saying "Ja" they were kindly treated, they decided that they ought to reply "Ja" to whatever they were asked.*

> *So it befell that when they were asked if they were heretics, and if they had come to Germany in order to infect it in the way they had perverted Lombardy [the Cathars], they replied "Ja." Whereupon some of them were beaten, some imprisoned, and others stripped and led naked to the local court, and made a sport for men to mock at . . . From this experience, Germany was considered by the Brethren to be such a ferocious country that only those inspired by a longing for martydom would dare return hither.*

In 1227, a new pope, Gregory IX, decided to eradicate what remained of the Albigensian heresy by appointing special prosecutors who had the authority to detain, question, and punish suspected heretics in the pope's name. For this he chose the Dominicans. They proved so successful in France that Gregory soon expanded the practice throughout Europe. So ferocious were the Friars Preachers in enforcing orthodoxy that a grim play on words circulated, separating *dominicanes* into *domini canes,* meaning "dogs of God." Thus was born the Inquisition, an institution that would sow terror in the name of God for the next 350 years.

In order to be effective in this and other worldly responsibilities, Dominic understood the necessity of having educated followers. Accordingly, seven of his original sixteen acolytes were sent to the University of Paris. Others went to Frederick's university at Naples, and by 1222 there were Dominicans at Oxford. The universities thus became de facto recruitment centers for the Friars Preachers.

Learning, on the other hand, was anathema to Francis, who wanted an order where all the brethren were completely equal. He worried that distinctions of birth, wealth, knowledge, or position would overcome his vision. As beloved as he was personally, however, Francis was unwilling to impose standards of behavior on his order, choosing to lead by example. Without a strong hand at the top, almost from the first, Franciscans followed the Dominicans to the universities.

Once at the university, to Francis's chagrin, the Friars Minor demonstrated the same zeal for learning (and recruitment of the learned into the order) that characterized their brother mendicants, the Friars Preachers. In fact, there was soon so much competition between the two for who could convert the most students that there were frequent complaints to the papacy that one or the other was engaging in unfair recruitment practices. By the time of St. Francis's death in 1224, most of his original vision had been modified or superseded by political practicalities.

IN 1228, DURING A CITYWIDE FESTIVAL IN PARIS, a brawl began in a bar and then spilled out onto the streets. Students were killed, and in protest, the masters shut down the university and vowed to leave Paris if their grievances were not addressed. They were not, and so there was a mass exodus of masters and students to other universities.

With the University of Paris closed, Roger Bacon's parents chose to send him to study closer to home at the new university at Oxford, which, unbeknownst to them, happened to be the most radical school in all of Europe.

CHAPTER FIVE

# Rebels in Gray Robes: Oxford

• • •

OXFORD WAS NOT EVEN A TOWN until the tenth century, when a wall was built as a defense against any invaders who might attempt to cross the Thames. There was no tradition of higher education—Oxford never even had a monastery school. In 1117, there is a record of one master and fifty pupils. Fifty years later, the small walled city in the rolling countryside along a strategic river became a favorite of Henry II (Richard and perhaps John had been born in nearby Beaumont Castle). By the beginning of the thirteenth century, Oxford was a thriving mercantile center.

Oxford became a university town for many of the same reasons that Paris did—scholars bearing translated manuscripts came there to settle and study. Enrollment got a big boost when Henry, furious that the French were providing sanctuary for Thomas à Becket, forbade English students from crossing the Channel to attend school.

As the number of students swelled, lodging and class space became harder to find and more expensive. Classes were held in a variety of public buildings, taverns, or church facilities, often wherever a master happened to be staying. The city was dangerous, particularly after dark when the gates were locked, with robberies and murders common. The young men who came to study, therefore, most just thirteen or fourteen years old, tried to procure lodgings close to the master with whom they would be studying.

Oxford was officially a church school, so anyone wishing to enroll was required to become a seminarian and shave a tonsure (a bald spot on the top of the head). Clerical status, however, did not prevent the young Oxford scholars from developing much the same relationships with the townsfolk as their counterparts in Paris. Many of the students, like the young Bacon, came from wealthy families, which did not endear them to a largely working-class populace. The tonsures made the students stand out, and they made it a point to speak to one another in Latin, incomprehensible to the locals, in effect saying, "We're better than you." Brawls were common, and tensions between the two groups often ran high as more and more people—to say nothing of livestock—were crammed within the city's walls.

Students roomed together to save money, and, in 1209, when one of them killed a local prostitute who was sharing his rooms, some citizens, led by the mayor, attacked and killed his two roommates in retaliation. The masters closed the university in protest and then migrated to other cities to teach—thus was Cambridge founded. Oxford stayed closed until 1214, when King John handed England over to the papacy and Innocent III issued a charter to the Oxford masters similar to the one that he had provided to the University of Paris.

By the time Roger Bacon entered the school in about 1228, Oxford had fully recovered and probably had more than a thousand students. As had happened in Paris, many students enrolled not to pursue knowledge for its own sake but because they knew that a university degree had become a necessary stepping-stone to a career in the Church or government. "A boy of parts goes to Oxford, let us say, with the help of a bishop or abbot or local landholder, or, as happened more frequently than is generally supposed, because he belongs to a family which can support him there," observed the Oxford historian Sir Maurice Powicke. "He makes good, and in due course incepts as master of arts. An influential teacher may open the way to a career . . . If he belongs to a well-to-do local family, the scholar's future is safe."

Oxford structured its arts curriculum along the same lines as Paris. It was no match in theology, but that turned out to be an advantage, since Paris, specifically because of the preeminence of its theology curriculum, remained under the close scrutiny of the pope. Not so Oxford. It was too far away and deemed not sufficiently important. As a result, by 1228, Oxford was able to deviate from generally accepted teaching principles. Not only did the school teach Aristotle, but it also placed more emphasis on mathematics and experimental science than any other university in Europe. That it did so was due almost entirely to the influence of one extraordinary man, Roger Bacon's spiritual mentor, Robert Grosseteste.

In an age where birth was everything, Grosseteste rose from poverty so extreme that no one is even sure of his family name. He was born around 1175 in Suffolk, and he must have demonstrated remarkable ability as a child. In 1192, probably through the auspices of a church patron, he was sent to study at Oxford. He so distinguished himself there that he was recommended to the bishop of Hereford and selected to study theology in Paris, an honor reserved almost exclusively for those of high birth. Afterward, he returned to teach at Oxford, eventually rising to become chancellor. In 1235, he was named bishop of Lincoln, another unheard-of honor for a man born little more than a serf, but he continued to oversee Oxford, which lay within his diocese.

Grosseteste was not simply a scholar and a cleric. He was the most influential English clergyman of his time. His passion for his office, and the high standards he tried to implement, reverberated throughout the country. When Bacon came to Oxford in 1228, Grosseteste had just taken on the additional role as advisor to the Friars Minor. Although he never joined the Franciscans himself, he had agreed to lecture to them, which provided an enormous boost in the prestige of the order at the university. Adam Marsh, a Grosseteste protégé and one of Roger Bacon's future teachers, a man who would become extremely influential in English politics, joined the Franciscans because of Grosseteste.

For all the political and ecclesiastic power he wielded, however, Robert Grosseteste's real legacy lay in his contribution to the advancement of science. "Master Robert, called Grosse-Teste, lately bishop of Lincoln, alone knew the sciences," Roger Bacon would later write, and:

> *For very illustrious men have been found, like Bishop Robert of Lincoln . . . who by the power of mathematics have learned to explain the causes of all things . . . Moreover, the sure proof of this matter is found in the writings of those men, as, for example, on impressions such as the rainbow, comets, generation of heat, investigation of localities on the earth and other matters, of which both theology and philosophy make use.*

Because of Grosseteste and Adam Marsh, Oxford became the center of scientific inquiry in northern Europe, and by association the order of itinerant, illiterate beggars envisioned by St. Francis became identified with the most forward-thinking intellectuals of the age.

Grosseteste himself was the junction from which all experimental science grew. He divided scientific inquiry into three levels, based on the degree of certainty with which a person could trust his findings. At the bottom was that which could be known with surety: mathematics, for example. That a triangle was a three-sided, closed figure was an objective truth. The next level Grosseteste called natural physics, which today would incorporate a broad range of natural sciences, including botany, zoology, astronomy, optics, and chemistry. Conclusions in natural physics could be drawn with some degree of certainty through intense observation, although an experiment must yield the same result after many different trials in order to allow the experimenter to hypothesize a conclusion. A sufficient number of such conclusions would eventually yield the presence of a natural law. The highest level was what the thirteenth century called metaphysics, which was the science of God and the soul. Scholars, Grosseteste said, could not prove truth in the realm of metaphysics with any degree of certainty—the study of God required divine inspiration.

Grosseteste employed a surprisingly modern form of scientific method. He carefully observed and applied mathematics to natural phenomena such as refracted light, the movement of heavenly bodies, thunder, the nature of clouds, or the physiology of horned animals, then deduced what he called a "definition" but what today would be called a theory. However, although he asserted the necessity of empirical verification, Grosseteste was not himself an experimenter.

When Roger Bacon enrolled at Oxford, he encountered this new dedication to science and mathematics. It is not clear whether he arrived early enough to actually attend Grosseteste's lectures, but by 1228 Grosseteste's spirit, academic commitment, and willingness to go where knowledge took him permeated the school. Aristotle's physics and metaphysics were on the curriculum, and there were classes in geometry and optics. It was an exciting time to be a student—there was the sense of breaking through academic barriers and advancing knowledge on a larger stage. "Here [at Oxford], mainly owing to Grosseteste's influence, the *libri naturales* [Aristotle's work on natural science] were early accepted and never had to go underground," noted the medieval historian Gordon Leff.

The masters, particularly Adam Marsh, recognized Bacon's potential immediately and accepted him into their inner circle. Forty years later, the pride in Bacon's voice was unmistakable when he discussed his part in that great intellectual movement. Describing the science of *perspectiva* (optics), he wrote to the pope: "However this science is not yet taught at Paris, nor among the Latins except twice at Oxford in England and there are not three people who know its power."

After he received his arts degree, the next step for someone as talented as Bacon would ordinarily have been the study of theology. With his credentials, he might well have been accepted at Paris, and the way thus paved for him to achieve great honors. But he had no taste for theology as it was taught in the thirteenth century. After experiencing the excitement and challenge of Aristotle, Avicenna, and Averroës, of *perspectiva* and the *Metaphysics,* he could not bring himself to the minute examination of the *Sentences,* which was still the core of the theological curriculum. He evidently considered it a waste of his time, and since in that age it took about thirteen years to achieve a theological degree, it was going to be a good deal of time to waste.

Even at this point in his life, still a young man, Bacon was already exhibiting the self-assurance, the conviction in the power of his own ideas, and the utter lack of political acumen that marked him as an outsider in an age of conformity. This willingness to trust in what he believed instead of what he was told or what everyone else was doing would define his life and his legacy. So instead of toiling away meaninglessly just to guarantee safe, respected professional advancement, he became a master of arts. Not only would this allow him to continue scientific research independently, but he could freely instruct others in that which he had studied himself.

There are those who claim that he later regretted this decision, that it caused him to become a bitter and brittle man. The argument goes that he threw himself into the study of science only for personal recognition and realized too late that fame could not be achieved without a theological degree. His whole idea of studying mathematics and the natural sciences through experimentation was just his way of becoming "a theologian by the back door," as one Bacon biographer, Stewart C. Easton, put it.

The rejection of the theological degree was not the great mistake of Bacon's life, however, but the making of it. Without the constricting rigor of the *Sentences,* he came at theology from a different perspective and saw that science was no threat to faith. Although he suffered for his nonconformity during his lifetime, he created a unique intellectual legacy, one that continues to resonate more than seven centuries later.

In 1237, now in his early twenties, Bacon began teaching Aristotle at Oxford. He might have stayed there in perpetuity, had the University of Paris not suddenly found itself in need of a master of arts to teach the *libri naturales*.

THE UNIVERSITY OF PARIS HAD REOPENED when Pope Gregory IX issued a new bull, *Parens scientiarum,* a charter of liberties that made clear the masters' privileges. The masters had accordingly returned to Paris, and the university once again opened its doors to students. But though the pope had made clear the masters' rights to decide what students could and could not wear, when landlords were charging too much for lodgings, and just how long the long vacation should be, he did not lift the ban on Aristotle. The arts masters chafed at the continued restriction; how could Paris be the leading academic institution in Europe and *not* teach Aristotle and the related Arabic commentaries? What was more, the *libri naturales* were still being read surreptitiously.

Gregory backed down and appointed a three-man committee to go through the *libri naturales* line by line, expunging any suspicious material. It was a huge job, and the three aging appointees—one died almost immediately—were not remotely up to it. Not a single page of the new sanitized Aristotle was ever issued. In the meantime, the unauthorized versions continued to make inroads, even into the theological faculty. Sometime in the late 1230s, the arts faculty felt sufficiently emboldened to send a request to Oxford for a master to teach Aristotle openly at the school.

This position was an enormous opportunity for whichever

Oxford master was chosen. Whoever went would have the chance to make a name for himself at the most prestigious institution of learning in Christendom. It was also an opportunity for Oxford to establish its credentials as an elite center of learning. The authorities at the school knew to send their best young master.

They sent Roger Bacon.

Of his stay in Paris, Bacon would write later, "I composed many things for the elementary instruction of youths." He was not much older than many of his students when he first arrived and was evidently very popular. Twelve of his lectures were published and have survived, and they describe lively class discussions with even some good-natured byplay between the students and the teacher. Master Bacon's charm was that of any good professor: he was approachable, enthusiastic, and engaging, a man who took his class members' arguments seriously and wasn't above admitting a mistake or learning from a discussion. His knowledge of Aristotle was superior but not infallible; he tells the story of mispronouncing a word that he thought was Arabic. "This word [was] not the scientific one but colloquial Spanish," he wrote. "After being laughed at by my Spanish students, familiar as they were with the words of their own language, when I did not understand what I was reading, I at length learned from them the meaning of this word and of many more besides."

Bacon lectured on everything from light and the nature of the universe to the descriptive qualities of the souls of vegetables. His geometry problems were the bane of his students' existence—he took great pride that not even the other arts professors could solve them. In class, he used the scholastic method, of course: a thesis would be posited from a work of Aristotle, to which an objection would be raised and answered, followed by another and another and another until one side convinced the other or a compromise solution was imposed by the master. For example, in a lecture on the philosophy of plants, the question was raised on what happens to the soul of a plant when it is grafted onto another. (This was evidently a tricky one. Bacon was forced to resort to compromise: unless the result of the graft was a new species, each plant retained its original soul inside the grafted plant.)

Perhaps the most difficult problem for Bacon was Aristotle's notion of the eternity of time and therefore of the world, which violated the scriptural provision that God created the world—and therefore time itself—in a finite moment. Because Bacon was a deeply religious man, there is almost a desperation in his attempts to reconcile Aristotle to Christianity without corrupting the Philosopher's words. "It is seen that he [Aristotle] has suggested nothing against the faith. For he says there will have been no motion when there was no time; so he only meant that motion did not begin in time, which is true. Aristotle argues elsewhere that there must be a first of everything or there would be no later things, so why not a beginning of the world?"

The genuineness of Bacon's give-and-take with his students forced him to refine his own thinking every bit as much as his students', and even to reconsider some of his solutions. The more Bacon studied, the more he saw how much there was to learn. Like present-day physicists searching for "the theory of everything," he decided that there must be one unifying science above the others, one grand concept that would reconcile nature, the scriptures, the universe, and the soul. And so by degrees there came over him what today would be considered a laughable idea.

He would learn everything there was to learn. He would master every science. And because he had now seen firsthand the advantages of understanding different languages, so as to read a text uncorrupted by a faulty translation, he enlarged his goal beyond the sciences. He would master the languages of Greek and Hebrew as well.

Bacon's decision to learn all that was knowable was not as unique or fanciful as it might seem. A Dominican who had recently arrived at the University of Paris was apparently already doing it.

CHAPTER SIX

# Science Goes Mainstream: The Rise of Albertus Magnus

• • •

HISTORY HAS ALMOST ALWAYS RESERVED the epithet "the great" for a king or conqueror. For Albertus Magnus that conquest was not of territory, riches, or rival princes, but rather of natural science. During the thirteenth century and long after, Albertus Magnus was considered perhaps the most accomplished scholar who had ever lived.

Albert was born in the town of Lauingen in Germany sometime around 1200—as with Bacon, there are no surviving records. The men in Albert's family had, for successive generations, been knights in the service of the Hohenstaufen emperor. His father had served Frederick II's father, Henry VI; his grandfather, Barbarossa. It was no doubt with the idea of perpetuating the family tradition that Albert was sent to Padua to live with his rich, well-connected uncle and to study arts at the new university there.

The University of Padua was founded by some wandering law professors and seems to have offered a basic arts curriculum and little else. Theology would not be taught at Padua until the fourteenth century, and there was no tradition of scientific inquiry as there was at Oxford. Even the study of canon law was neglected. Moreover, Albert's tenure as a student predated the distribution of the new translations of Aristotle, so he would have been limited to the standard logic texts. He did, however, display an acute interest in nature and medicine, which in those days revolved largely around the study of plants.

Then, in 1223, while he was still young enough to be under the supervision of his uncle, the Dominicans came to Padua.

It was the early days of the Friars Preachers, and there was a heady aura of piety, mystery, and commitment surrounding the order. The friars made no secret of their pursuit of educated recruits. Jordan of Saxony, the immediate successor to Dominic himself, came to entice members of the student body at Padua into the order. Jordan already had a reputation for charismatic preaching: "It is said of him that he drew out of the stormy sea of the world, with the net of the Divine Word, and clothed with the white tunic of his newly born Order, more than a thousand young men belonging to the Universities of Paris and Bologna," wrote a medieval chronicler. After hearing him speak, ten Padua students joined on the spot, "among them the sons of two great German lords; one was a provost-marshal, loaded with many honors and possessed of great riches; the other has resigned rich benefices and is truly noble in mind and body," Jordan later boasted. This second reference was to Albert.

By all accounts the family was appalled; the uncle intervened, claiming that his nephew was too young to commit to the order, and dragged him home. It did not work; by Easter, Albert had shaved his tonsure and donned the traditional coarse clothing of the Friars Preachers. Some twenty years later, a young Thomas Aquinas would be recruited to the Dominican order while studying at the University of Naples in much the same way—and also to the dismay of his family—prompting Roger Bacon to later write derisively, "These are the boys among the students of the two orders like Albert and Thomas, and others, who entered the orders when for the most part they are twenty years of age and less."

Soon after converting, Albert was sent to the Dominican priory at Cologne to study the Bible. They'd never had anyone like him there before—in 1225 he began studying theology with the approved lector, and by 1228 he *was* the approved lector. These years in Cologne were crucial to Albert's development. He was at the height of his energy and intellectual powers, and the administrative and political duties that would claim so much of his attention in later life were relatively light. He could therefore devote himself to his reading and especially his study of nature.

Although he was eventually to write and lecture on nearly every conceivable subject in the medieval curriculum—alchemy, astronomy, astrology, math, theology—Albert was primarily a naturalist. He learned the names and characteristics of hundreds of plants and minerals. It was while stationed at Cologne that he would go miles out of his way to visit mining towns in order to locate precious metals and rock specimens. In his leisure hours he stared up at the sky and once was rewarded by the appearance of a comet.

With the conversion of Albert, the Dominicans had acquired an asset far beyond their expectations. Albert's devotion to learning was matched only by his devotion to his order. He was to win a public recognition that not only reflected favorably upon the Friars Preachers but could be translated into tangible political gains. All his life, as he tramped from city to city at the behest of first his superiors in the order and later the pope (for Albert walked everywhere, even in old age, although he had a donkey for his precious books), he indulged his passion for study. "It was his custom while traveling . . . first to visit the chapel of the religious house where he intended to stay the night, to thank God for the safe journey, then immediately to visit the library to see whether there were any books that he had not yet seen," reported the Dominican scholar James Weisheipl. Over the years, he was sent all over Europe and had the chance to observe and learn in many different environments. As a result, he accumulated knowledge that for the time approached the encyclopedic, earning him the nickname *Doctor Universalis*.

But although his scientific learning was prodigious, Albert's theological training was virtually nonexistent. As a naturalist, he identified with Aristotle, or at any rate with whatever Aristotle he had managed to pick up along the way. But of the new translations circulating in places like Oxford and Paris, he was ignorant. Books like Michael Scot's simply didn't reach towns like Cologne in the 1230s. Neither did Albert have the advantage of disputing or conversing with other scholars—in Cologne, he *was* the scholar. Accordingly, when it came to theology, Albert was almost completely self-taught.

Nonetheless, word of his superior intellect and knowledge spread through the order. When the opportunity arose for a Dominican friar to become a master of theology at the University of Paris in 1244, the master general selected the supremely unqualified Albert for the job. Nothing bespeaks the rising power of the mendicant orders in the first half of the thirteenth century better than their success in muscling a brother with no training into the theological faculty at Paris. The implications went far beyond the simple question of who would teach whom. There were vast political stakes as well. Control of the theological faculty at Paris, with its unique role of training future bishops, cardinals, and even popes, might easily be translated into control of the Church.

THE TWO ORDERS FIRST CAME TO PARIS specifically to recruit learned members. The friars had no students in the undergraduate arts courses; mendicants were forbidden from attending secular arts schools. Instead, they had set up their own undergraduate schools to teach what they thought was appropriate. Needless to say, Aristotle and the new translations were not part of the curriculum.

In theology, however, even the friars recognized that the faculty of the University of Paris was so far advanced that there was no choice but to take advantage of its teaching. Accordingly, both the Dominicans and Franciscans selected some of their more promising members to study theology there. By good luck (or calculation, as was later accused), the first of the mendicants to qualify as a theology master, a Dominican by the name of Roland of Cremona, did so just before the university disbanded in 1229.

Roland, being a friar, did not consider himself bound by his degree to act in concert with the other theological masters, so he stayed in Paris with his order to teach the brothers what he himself had just learned. When word spread that there was a master teaching at the Dominican school, a number of theological students who had declined to follow their masters out of town asked if they could listen in.

For the Dominicans, this was a chance to insinuate themselves even further into their surroundings, to show off the advantages of the order and maybe induce a few more young men to join. Roland of Cremona was allowed to open his classes to secular students.

Rarely has such a simple act had such far-reaching repercussions. Roland was evidently a popular teacher, so that when the University of Paris reconvened in 1231, these new students stuck with him. The theological faculty was faced with a *fait accompli*: a Dominican now seemed to hold one of the all-important chairs in theology. The secular masters were still fuming about this when two of their own, John of St. Giles, the master who had taught Roland, and Alexander of Hales, possibly the most important master on the entire faculty, also converted. John went to the Dominicans and Alexander to the Franciscans. Suddenly, the mendicant orders held three of the twelve theological chairs at the University of Paris.

The secular masters, lay professors who viewed the university as a place of lifetime employment and who had no other means of earning their livelihoods except by collecting fees from students, were furious. It was dangerous, they said. There was no way of controlling what was being taught in other places once the mendicants left. To make it worse, the mendicants were extremely popular teachers, using poverty and piety to lure students away from the secular professors. Losing students meant losing income, which to the seculars, who were not supported by an organization, was a matter of survival.

Until the friars came into the picture, no one who was granted a chair in theology would think of leaving the university. It was a theology master's job to stay on in Paris to teach and help run the school. But the mendicant masters had no allegiance whatever to the university and so stayed only a short time—some for less than two years—before reassignment to a Franciscan or Dominican school in another country to spread the learning. A new candidate would then take the place of the friar who was leaving. And since once acquired a chair held by a Dominican or Franciscan would be passed on to a Dominican or Franciscan, with just twelve chairs at stake, each became a symbol and measure of growing—or declining—political power.

Moreover, in their rush to push people through, the orders took shortcuts. Those applying for the Franciscan or Dominican chairs were in their forties and fifties, men who had held positions of authority within their orders before coming to Paris. Surely, claimed the mendicants, men like these should not have to spend additional years of their lives reading and lecturing on Peter Lombard's *Sentences* the way a twenty- or thirty-year-old did? After all, they had already lectured on the Bible to members of their own orders. The friars were thus able to reduce the requirement for a chair in theology from an average of thirteen years to sometimes as little as two or three.

Thus was Albertus Magnus granted a chair in theology in Paris after just two years of study. He arrived at the university in 1243, was incepted as a master in 1245, and then relinquished his chair in 1248 to the next Dominican.

Not that Albert did not try to justify his position—during his five-year stay in Paris, he began not one but two massive projects. The first was a commentary on Peter Lombard's *Sentences,* which grew out of his own lectures and eventually ran to seven volumes, and the second was his own *Summa Parisiensis*, a written account of all of his public disputations at the university. In addition to these two written works, he lectured to theological students, learned what he could of the new translations of Aristotle and the related Arabic commentaries, and handled his share of university responsibilities, which in 1248 included adding his signature to a decree condemning the Talmud as heretical and ordering that the books be burned publicly.

Roger Bacon, watching all of this from his seat on the arts faculty, was appalled. How could Albert earn a chair in theology in just two years when it had taken Robert Grosseteste and Adam Marsh—far greater men than Albert in Bacon's opinion—more than a decade? And Albert had done all of this without once having to attend an arts class of the kind that Bacon taught and took so seriously. Albert knew no Greek, no Hebrew. What did the Dominican know, what *could* he know after such a short time, of *perspectiva*, of mathematics, especially of geometry, the crux of Bacon's own education and at the core of his belief of how the world worked?

Twenty years later, Bacon would give voice to these objections in a treatise to the pope. He did not mention Albertus Magnus by name, but it is clearly he who is described:

> *The . . . one who lives [Albertus Magnus] entered an Order of Friars as a boy. He never taught philosophy anywhere, nor did he hear it in the schools, nor was he in a* studium solemne *before he was a theologian, nor was he capable of being taught in his own order, as he was the first master of philosophy among them. And he taught others; whence from his own study he had what he knows. And truly I praise him more than all of the common students, because he is a most studious man, and he saw many things, and had money. And so he was able to collect many useful things in the infinite sea of authors. But since he did not have a foundation, for he was not instructed or exercised in hearing, reading, or disputing, it was inevitable, therefore,*

> *that he did not know the common sciences. And again, since he did not know the languages, it is not possible that he would know anything great, on account of the reasons which I write concerning the knowledge of languages. And again, since he ignores perspective, just as others of the common students do not know it, it is impossible that he should know anything of worth about philosophy . . . God, however, knows that I have only exposed the ignorance of these men on account of the truth of study. For the common student believes that they [Albertus Magnus and Alexander of Hales, whom Bacon also criticized] know everything and they adhere to them like angels . . . And especially the one who lives [Albertus Magnus]; he has the name of* doctor Parisius.

Albert seems to have been aware of Bacon and his criticisms. At one point, he wrote of "certain indolent people, who looking for comfort in their indolence, look at nothing in script except what they can criticize."

Albert became even more of a nemesis to Bacon after he left Paris. He had returned to Cologne by order of the minister general to establish a new university. He set the tone for this new school as Grosseteste had once done for Oxford. Albert, who had come to the new translations only a few years before, now undertook the greatest compilation and analysis of the works of Aristotle ever seen in the Christian world. "Our purpose in natural science," he wrote, "is to satisfy as far as we can those brethren of our Order who for many years now have begged us to compose for them a book on physics in which they might have a complete exposition of natural science and from which also they might be able to understand correctly the books of Aristotle."

*Albertus Magnus* EDGAR FAHS SMITH COLLECTION, UNIVERSITY OF PENNSYLVANIA LIBRARY

Albert attacked the task with painstaking deliberation. In page after page, book after book, he explained Aristotle, carefully mirroring the Philosopher's own works, such as the *Physics, De mineralibus, De Anima, De vegetabilibus, De animalibus*. He covered general principles and universals, changes in animate and inanimate objects, the human soul, the animal soul, the soul of plants, perception, imagination, instinct, memory. He paraphrased when necessary, often adding his own interpretations to the stickier problems and filling in gaps wherever he felt it necessary. Where Aristotle deviated from Christian dogma, Albert simply said that Aristotle was wrong.

It was this encyclopedic work on Aristotle that made Albertus Magnus famous across Europe. Because his books were intended for use in Dominican schools, they went everywhere. As a result, even though Albert's work lacked critical analysis or depth, it was he and not Roger Bacon who became known as the foremost expert on Aristotle in Christendom, the man most responsible for bringing the Philosopher's work into the classroom.

But Albert's commentary and analyses did more than promote his own fame. Unlike the work of Grosseteste, who had stayed in a provincial part of provincial England and who never got into the mainstream, Albert's work helped validate science as a legitimate branch of theological study.

Although both Albert and Bacon each believed that the other was engaged in a perversion of science, they were actually not that far apart in either method or outlook. Both undertook to explain the natural world by extrapolating from sensory observation to more general principles. Bacon was theoretic—more Newtonian—while Albert was a scrupulous observer—more Darwinian. If not for the politics at Paris and, later, rivalry between the orders, the two might have ended up as allies. As it was, the only real source of argument between them was the importance of mathematics to scientific theory. Bacon believed that mathematics underlay everything in nature, and Albert rejected this, although, as was true on many other occasions, he offered no proof or underlying reasoning for his conclusion:

> *[It is an] error of Plato [and of others, like Bacon] . . . who held that natural things are founded on mathematical and mathematical being formed on divine, just as the third cause is dependent on the second, and the second on the first; and so [Plato] said that the principles of natural being are mathematical, which is completely false.*

Albertus Magnus would remain Bacon's enemy until the day he died, but it was Albert's protégé, Thomas Aquinas, and his rejection of experimental science that would bring about the ruin of Roger Bacon.

CHAPTER SEVEN

# The Dumb Ox: Thomas Aquinas

• • •

ST. THOMAS AQUINAS, one of the most revered names in all of Catholic history, was born Tommaso d'Aquino in 1224 or 1225 in the ancestral castle of Roccasecca, about fifty miles north of Naples, in what was then the kingdom of Sicily. He was the youngest child and only son of his father's second wife, a member of a noble family allied closely to Frederick II. Many of Thomas's uncles and cousins had served the emperor, and his half brothers had been pages at Frederick's court and later soldiers in the imperial army. Although physically, temperamentally, and spiritually no two men could be more different than Frederick and Thomas, they are inexorably linked, both by fortune and coincidence.

In 1229, when Thomas was four or five years old, Frederick had pulled off the most dazzling coup of an astonishing life. He had long been in contact with al-Kamil, the sultan of Egypt. The two had exchanged geometry problems and gifts, and over time they had become familiar with each other's political realities and ambitions. Frederick wanted Jerusalem, al-Kamil wanted Syria, and they worked out a deal of mutual support. The result was a ten-year truce with powerful Egypt, which included the return of Jerusalem to Christendom. Thus Frederick achieved without the loss of a single human life what hundreds of thousands across Europe had suffered and died for since the very first crusade had been called in 1095.

Frederick rode into the Holy City on March 17, 1229, and had himself crowned the king of Jerusalem. It had been Thomas's father's cousin, another Tommaso d'Aquino, who had handled the delicate negotiations with al-Kamil.

There was a tradition in Sicily at the time that the youngest son in a family was given to the Church, and so, at almost the same moment as the emperor was riding into Jerusalem, Thomas's parents sent him off for schooling to the Benedictine abbey of Monte Cassino, just a few miles to the north. The Benedictines were aligned more with the pope than with the emperor, but that did not seem particularly inflammatory at the time. What's more, the Aquino family was sufficiently prestigious (and had donated enough to the abbey) to be fairly confident that Thomas would one day be made abbot.

Monte Cassino was not much of a school, but it sufficed for primary education. Thomas learned a little reading, a little writing, and a little Latin, along with the psalms. He probably would have stayed on and become a Benedictine monk himself had the further intrigues of Frederick II not intervened.

Frederick's ambitions were not limited to the Holy Land. He wanted Italy as well, particularly Rome itself. He and Pope Gregory IX—the current ruler in the city—loathed each other with single-minded fury. Gregory had long since excommunicated Frederick and had tried to seize Sicily while the emperor was in Jerusalem by starting a rumor that Frederick was dead. Frederick foiled the plan by turning up alive, and for the next decade the two fought for the future of Europe and the Church.

They alternated between a war of words and one of deeds. Frederick called the pope "the abomination of Babylon," while Gregory spread the word that the emperor was the Antichrist. At the end of the twelfth century, a Calabrian monk named Joachim of Flora had prophesied that a representative of Satan was to bring about the end of the world, the Day of Judgment, and a new era in or around the year 1260. Joachim's prophesies had spread throughout Europe, gaining adherents throughout the Church, particularly among the Franciscans and others prone to mysticism. Gregory simply attached Frederick to Joachim's prophesy. "With fangs and claws of iron it seeks to destroy everything and trample the world to fragments beneath its feet . . . behold the head and tail and body of the Beast, of this Frederick, this so-called emperor," Gregory noted in a proclamation.

By 1236, in one of the more active periods of this conflict, Frederick had decided to take Lombardy, and full-scale war ensued. There was a decisive battle at Cortenuova, which Frederick won by skill and daring. Within the week he had swept into the principal city of Cremona in full regalia, his prisoners trailing behind him, the nobles in chains. The foremost prisoner, a Milanese aristocrat, was tied to the mast of a captured Milanese vessel, which was pulled through the streets by Frederick's ever-present elephant. The man chosen to oversee the guarding of these prisoners was Thomas's father. His father's cousin, the other Tommaso d'Aquino, was Frederick's envoy in Rome.

Eventually, in 1239, imperial troops took over the abbey of Monte Cassino, and it was suddenly inconvenient for a member of the Aquino family to be a novitiate at a facility even loosely aligned with the pope. Thomas left and came home.

Thomas was at just the right age to go to university, so his parents sent him off almost at once to the new school Frederick had established at Naples. Not surprisingly, Frederick's university was like no other in Europe. Unlike the schools at Paris, Oxford, and Bologna, which had grown out of a tradition of scholarship, Naples had been created by imperial fiat for the express purpose of finding talented boys and training them to be civil servants. In those days, this meant the study of law. Frederick even appointed his own chief judge as a professor and paid other qualified legal professionals to teach. It was the only school in Europe funded not by the students but by the state.

The arts program in which Thomas enrolled was therefore heavily weighted to law, which became his first discipline, just as math and science had been Bacon's. Because this was Frederick's school, however, it also taught the *libri naturales* and the related Arabic commentaries—Michael Scot's translations were available in Naples years before they came to Paris—and Aquinas studied these as well. But he would always come at natural philosophy from a lawyer's point of view, and it does not appear that he had much if any mathematical training. In 1241, midway through Thomas's stay in Naples, Gregory IX died just as Frederick's armies were closing in on Rome.

Thomas stayed at the University of Naples until 1244, long enough to incept as a master of arts, although he never actually earned his degree. Instead, like Albertus Magnus twenty years before him, he fell under the spell of the Friars Preachers.

A deeply religious man, Aquinas had never given up the notion of going into the Church, but the Benedictine order was not what he was looking for. Although the studious side appealed to him, it was too removed from participation in Church and social policy. Years later, he would write in his *Summa Theologica*, "the highest place among religious orders is held by those which are ordained to teaching and preaching, which functions belong to and participate in the perfection of bishops." "Teaching and preaching"—that was certainly the Dominicans. They had been in Naples since 1227, no doubt lured by the presence of the university. In 1231 they were given a church and priory, and it was here, in 1244, that the nineteen-year-old Thomas entered the order.

The Dominicans were naturally pleased with this new aristocratic recruit, but having an Aquino around the priory was a mixed blessing. The Friars Preachers were by now known to be the pope's men and thus only barely tolerated by Frederick. The order had already had one irate noble Italian family aligned with the emperor break in and take back a son who had pledged to become a Dominican novitiate.

The friars decided that the best course of action would be to get Thomas out of town, so they assigned him to the University of Paris. The master general of the order himself, John of Wildeshausen, the man who had sent Albertus Magnus to Paris to take the Dominican chair at the university, undertook to escort Thomas to France. Naturally, they walked.

It is here that great events of state collided with a small family drama to produce a saint.

. . .

AFTER GREGORY DIED IN 1241, the civil authorities in Rome understood the urgent need for a papal successor. To ensure the prompt participation of the cardinal-electors, they had them rounded up by soldiers and harried through the streets to the meeting place, one older cardinal apparently dragged by his hair.

The election was headquartered in a ruin of a building, and all ten cardinals were herded into a small suite of rooms and locked in. The soldiers were lodged in the rooms directly upstairs. The structure was so decayed that the floors were rotting, which was convenient for the soldiers, since they could use the holes in the floor as toilets. This made the cardinals' bedroom in effect the soldiers' bathroom. Add to these foul conditions the sweltering heat of August, and it should come as no surprise that all ten cardinals fell ill and three died.

Despite what must have been a strong incentive toward a swift choice, it took two months to pick Gregory's successor. Finally the remaining cardinals chose one of their own, but his rule lasted exactly seventeen days before he died from the fever he had caught while deliberating in his own election. By that time, the other cardinal-electors had fled Rome. They weren't going through *that* again. It would be 1243 before Gregory's successor would be enthroned. Still, the interregnum had its advantages—rather than continue battle, Frederick waited to see who would inherit the papacy.

The new pope—who, tellingly, chose the title Innocent IV—demanded the return of all of the papal territory that Frederick had conquered while Gregory was alive. Frederick refused but asked to negotiate. Innocent agreed to meet him in Terni, north of Rome, in August 1244. Where Frederick went, so went his army, and in his army was Reginald d'Aquino, Thomas's older brother. Reginald had gotten wind from the family that Thomas was walking to Paris in the company of the master general of the Dominicans. By August they had made it as far as Terni.

What happened next was written down by Tolomeo of Lucca, a chronicler and early biographer of St. Thomas:

> *No sooner had Reginald heard that his brother was in the neighborhood (Frederick meanwhile pretending not to know what was about to happen) than he took Peter of Vineis with him and some men at arms, and went and violently separated his brother from the master general, and forcing him to mount a horse, sent him off with a strong guard to one of the family castles in the Campagna calle San Giovanni.*

The soldiers first used a manly approach to dissuading Thomas from his new brethren: they shut him in a room with what James Weisheipl described as "a ravishing girl, seductively attired." Thomas sent her away by brandishing a flaming stick from the fire. When he hurled it at the fleeing maiden, the story goes, the torch made the sign of the cross on the bedroom door. Thomas considered this a sign from God. Voluptuous farm girls failing, Reginald then handed Thomas over to his mother and sister for further persuasion (his father had died the year before), but it was Thomas who succeeded in persuading his sister to take orders and become a Benedictine nun.

Negotiations between Frederick and Innocent failed, and the pope escaped to Lyon. There, beyond Frederick's reach, he called a council and declared the emperor officially deposed. The Aquino clan, hedging its bets, then decided that perhaps having a Dominican in the family wasn't such a bad idea after all. Thomas's mother accordingly allowed him to "escape" (he climbed out a window of the family castle), and in 1245 he finally arrived in Paris.

AS A STUDENT IN PARIS, Thomas was quiet and without distinction. Albert doesn't even mention knowing him, although clearly he must have since the two of them left together for Cologne in 1248. Even at Cologne, Thomas, a heavyset man with large, sad eyes, spoke so little that his classmates called him *Boven Mutum*—"Dumb Ox."

Then one day a fellow student offered to help Thomas with his homework and wound up getting a lesson himself. He then begged Aquinas to tutor *him*. A little later, another student picked up some notes that Aquinas had dropped on the floor after class and showed them to Albert. Albert was so impressed with the line of argument that he called on Thomas at the very next session to respond to a particularly dense bit of scholastic reasoning. Aquinas's performance was brilliant, and from that day on he became Albert's protégé.

Thomas and Albert were an unusual pair. Albert was much more absorbed in the natural world and had the advantage in observation and experience. But Aquinas had the benefit of six years of arts study at Naples, which included all that intensive legal training, and was probably more familiar with the theological implications of the *libri naturales* and the Michael Scot translations than his master. Both men recognized the threat to the Church from the ideas put forward in the new translations of Aristotle. Albert, who would go on to many high administrative positions, quite possibly recognized in Aquinas's approach a theoretic sophistication which he himself was unable to provide.

Whatever his motive, so impressed was Albert with Aquinas's potential that when John of Wildeshausen asked him in 1252 who he thought would be a good candidate to study theology at Paris and take over one of the two Dominican chairs, Albert immediately recommended Aquinas, even though at twenty-seven he was much too young for the position. John of Wildeshausen protested, but Albert insisted, and off went Thomas Aquinas to Paris.

"We call him the Dumb Ox," said Albert, "but one day the bellowing of that ox will resound throughout the whole world."

ROGER BACON WOULD NOT HAVE MET AQUINAS when Thomas returned to Paris in 1252. Sometime around 1247, Bacon made the decision to reject Paris and everything it stood for—the politics, the close-minded rigidity of the theological faculty, the absence of that intangible spirit of discovery that had characterized his earlier studies. He chose instead to seek knowledge on his own. He left not to end a chapter of his life but to begin one.

It would be the great quest of his soul, this pilgrimage to truth through a comprehensive knowledge of science. It would occupy the rest of his life. It would demand all of his energy and resources, all of his faith and the faith of his friends. It would raise him up to great heights of intellectual and spiritual enlightenment and cast him down into the depths of intolerance. He would be both celebrated and despised as a result of the simple act of doing everything he could to learn, independent of conventional channels of education.

***Thomas Aquinas gestures toward a mountain in which sulfur and mercury combine to form metals, in a seventeenth-century alchemy text*** EDGAR FAHS SMITH COLLECTION, UNIVERSITY OF PENNSYLVANIA LIBRARY

To begin his quest he went back to the place where he had first encountered the excitement of science, the place he would return to over and over in the course of his life, the place where he felt most at home, from which he drew strength and vision.

He went back to Oxford.

CHAPTER EIGHT

# The Miraculous Doctor: Roger Bacon at Oxford

. . .

AT OXFORD, in addition to teaching at the university, Bacon threw himself into research, returning to his original interests in languages, mathematics, optics, and experiment. He bought books or had them copied whenever he could, always seeking rare editions in the original languages. Among the authors he would later cite were Ptolemy, Pliny, Seneca, Hali, Solinus, Tullius, Artephius, the Venerable Bede, and St. Augustine—this, in addition to Aristotle, Plato, Avicenna, Averroës, and the scriptures.

For the next ten years, he traveled to discourse with scholars in other towns and kept up a line of communication with those he most respected, such as William of Auvergne, John of Basingstoke, and a mysterious scholar named Peter of Maricourt, also known as Peter Perigrinus. If Bacon had a role model other than Grosseteste, it was Peter. Bacon later wrote glowingly of this man, observing that he refused all honors in order to be free to experiment and advance his knowledge. Almost nothing is known of Peter, and all his work has been lost except for a single paper, one of the earliest studies on the effects of magnetism.

Bacon also counted Adam Marsh among his circle of academic confidants. Marsh was lecturing in theology at Oxford as late as 1250, and Bacon probably attended his lectures. "I have sought out the friendship of all wise men among the Latins," he would later write. Grosseteste was still alive, and although he had many duties as bishop of Lincoln, he was still thoroughly involved in scholarship, learning Greek and producing a new translation of Aristotle's *Ethics*. Although there is no direct evidence that Bacon met or wrote to Grosseteste during this period, they were certainly aware of each other's work.

In addition to teaching, studying, and traveling, Bacon also conducted experiments with light, magnetism, and metals. It was a daunting schedule, and Bacon became renowned for working eighteen hours a day. "Men used to wonder . . . that I lived owing to my excessive labors," he would write later to the pope. He spent an enormous sum on instruments and materials for his work. "During the twenty years in which I have labored specially in the study of wisdom, after abandoning the usual methods, I have spent more than £2000 [a fortune in those days] on secret books and various experiments, and languages, and instruments, and mathematical tables, etc.," he wrote in the 1260s.

Bacon did not leave a lab book or notes, so it is impossible to re-create his actual experiments over the next twenty years. All we have are the stories that grew up around him. It was said that he astonished and frightened his students by directing light onto a crystal, thereby producing a rainbow, a feat generally attributed in the medieval mind to God alone. He was rumored to have produced explosions louder than thunder, which shook his study in the dead of night, and owned a magic looking glass that allowed him to see for fifty miles in any direction. One of his most picturesque supposed inventions was chronicled in a sixteenth-century potboiler entitled "The Famous Historie of Fryer Bacon, containing the Wonderful Things that he did in his Life, also the Manner of his Death, with the Lives of the two Conjurers, Bungye and Vandermast." Here he was credited with constructing a brazen or talking head (although a fourteenth-century tract attributes this achievement to Grosseteste).

> *Fryer Bacon reading one day of the many conquests of England, bethought himselfe how he might keepe it hereafter from the like conquests, and so make himselfe famous hereafter to all posterities. This (after great study) hee found could be no way so well done as one; which was to make a head of brasse, and if he could make this head to speake (and heare it when it speakes) then might hee be able to wall all England about with brasse. To this purpose hee got one Fryer Bungey to assist him, who was a great scholler and a magician (but not to bee compared to Fryer Bacon); these two with great study and paines so framed a head of brasse, that in the inward parts thereof there was all things like as in a naturall mans head.*

It was during this period that Roger Bacon acquired the title *Doctor Mirabilus*, or the Miraculous Doctor.

On a less fanciful level, Bacon was by all accounts a superb teacher. His students adored him, even though he occasionally shook them up with the crystal experiment or the demonstrations of gunpowder. He was even known to play a joke now and then. There is a story of a time when some Cambridge students descended on Oxford, intent on besting their rivals in a public dispu-

*Roger Bacon balancing the four elements, earth, air, fire, and water, from a seventeenth-century alchemy text* EDGAR FAHS SMITH COLLECTION, UNIVERSITY OF PENNSYLVANIA LIBRARY

tation. Bacon, forewarned of their approach, disguised himself as a laborer and met them just outside of town. When the students asked for directions, he answered them in Latin and challenged them to an on-the-spot debate. Astonished at the apparent level of education of a mere peasant, the Cambridge students turned tail and fled for home.

A LONG LETTER OF BACON'S that gives some idea of where his scholarship was directed and how much he had already accomplished has survived from this period. Entitled "The Letter of Roger Bacon Concerning the Marvelous Power of Art and of Nature and Concerning the Nullity of Magic," it is thought to have been written about 1248. There is no address, but it is clearly a reply to a knowledgeable correspondent.

In this letter, Bacon scorned sorcery and proclaimed that knowledge of nature, combined with knowledge of the way to use nature to reach practical ends, is a much more potent force. This disavowal is significant because some later writers have suggested that Bacon himself was merely a superstitious crank who believed in the black arts. Quite the contrary: Bacon declared that most of what the world called magic was merely sleight of hand, "wonders which do not have the truth of existence." He went through the various aspects of magic—invocation of spirits, magic characters, figures and charms—one by one and pointed out the flaws (or fraud) of each.

The real value of the letter, however, was in Bacon's assessment of what was possible through an acute understanding of the natural world, "marvels wrought through the agency of Art and of Nature . . . [in which] there is no magic whatsoever, because, as it has been said, all magical power is inferior to these works and incompetent to accomplish them." He wrote of the possibility of enormous ships propelled at amazing speeds without a single oar, of cars that move without benefit of horses or oxen, of engines, and even of flying machines. He wrote of bridges that span vast rivers without supports. "It is possible also that devices can be made whereby, without bodily danger, a man may walk on the bottom of the sea or of a river," he wrote, and "devices may be so contrived that the largest objects appear smallest, that the highest appear low and infamous, and that hidden things appear manifest." He talked of building strategic weapons that would emit poison and infectious diseases; he even mentioned "that device by which rays of light are led into any place that we wish and are brought together by refractions and reflections in such fashion that anything is burned which is placed there."

He reported on the alloying of gold and the ability to prolong life through regular exercise, moderate diet, and adequate sleep. His greatest hope was for a working astronomical model, in which "all heavenly bodies are described veraciously as regards longitude and latitude."

> *A faithful and magnificent experimenter might aspire to construct an instrument of such materials and of such an arrangement that it would move naturally in the diurnal motion of the heavens, a thing which seems possible because many things are determined by the movement of the heavens, such as comets, and the tides of the sea, and other things wholly and in part. In the presence of this instrument all other apparatus of the Astrologers, whether the product of wisdom or mere vulgar equipment, would cease to count any more. The treasure of a king would scarcely merit comparison with it.*

There is nothing in medieval science to compare to this one letter of Roger Bacon's. It is a masterpiece of insight and ingenuity; a thirteenth-century thought experiment at a time when science had not progressed much further than superficial observation. As a statement of purpose, it put him on the highest plane of scientific research and demonstrated the breadth of his understanding. Most of all, it vindicated the eccentricity of his approach.

The final two sections of this letter have been of enormous interest to Bacon scholars and the subject of great controversy for centuries. The latter consisted of a number of different recipes to produce what was known as the "Philosopher's Egg," a mystical material for turning base metals into gold. The recipes are detailed but completely incomprehensible, sometimes descending almost to gibberish. Some have theorized that these pages were not written by Bacon at all but appended afterward by some unknown author, although there was absolutely no evidence to support this conclusion.

Eventually, the section immediately preceding the Philosopher's Egg recipes came to be seen as a clue. Here, Bacon wrote on the "Wisdom of Keeping Secrets." "A man is crazy," Bacon wrote, "who writes a secret unless he conceals it from the crowd and leaves it so that it can be understood only by effort of the studious and wise." Then Bacon expounded on the "Seven Ways of Concealing Secrets." Codes and ciphers had been used since the Greeks, of course, but according to Simon Singh in *The Code Book*, Bacon's letter was "the first known European [work] to describe the use of cryptography."*

Bacon's seven methods include hiding a message "under characters and symbols" (a code); "in enigmatical and figurative expressions" (also a code); by "a method of writing, as by writing with consonants only like the Hebrews, Chaldeans, Syrians, and Arabians, and as the Greeks do" (shorthand ciphers); "by intermixing various kinds of letters" (cipher with nulls, meaningless characters inserted in a message to add a layer of complexity); "by means of special letters, devised by their own ingenuity and will,

* To quote Singh, "Technically, a *code* is defined as substitution at the level of words or phrases, whereas a *cipher* is defined as substitution at the level of letters." Both fall under the larger heading of *cryptography,* which is to hide the meaning of a message, and which, in turn, is a subset of *steganographics,* which is to hide its very existence. Obviously, any code or cipher that succeeded in obscuring the very existence of a hidden message would be far more effective than one in which the existence of something secret was obvious.

and different from those anywhere in use" (possibly artificial language); "actual letters are not used but other geometric figures that function as letters according to the arrangement of points and marks" (a code); and finally, "still a better way of obscuring which is comprehended in the *ars notaria* which is the art of noting and writing with whatever brevity we wish" (cipher in shorthand). Then Bacon added, "I have judged it necessary to touch upon these ways of concealment in order that I may help you as much as I can. Perhaps I shall make use of certain of them because of the magnitude of our secrets."

Bacon was a student of cryptography as well as a practitioner. He gave historical examples of the seven methods, and earlier in the letter he had written about magic symbols: "Certain of these irrational inscriptions have been written by philosophers in their works about Nature and about Art for the purpose of hiding a secret from the unworthy." In other words, much of what the common man considered a magic spell was simply a particular scholar's code, meant to protect a scientific secret.

There were those who believed that when Bacon wrote, "perhaps I shall make use of certain of them," he meant that he would do so immediately following that sentence. In 1914, a retired British colonel and amateur cryptographer, H. W. L. Hime, suggested that the Philosopher's Egg recipes were actually a secret message, employing what was known as "the Argyle cipher" to disguise the true formula of a substance that Bacon felt should not fall into the wrong hands. The Argyle cipher, known in the thirteenth century, was produced by laying a template with specifically placed cutouts over a page of writing so that only certain words remained uncovered. The same template could be made for an entire document or switched page to page, thus making it almost impossible to decipher without the correct key. An example of the Argyle cipher method had been included by William Makepeace Thackeray in his novel *Henry Esmond.*

When Colonel Hime laid his makeshift template across the last pages of Bacon's letter, what was left was a clear and accurate formula for making gunpowder, the first of its kind to be published in Europe. Whether Bacon had invented the substance or accessed the formula from one of his many readings of exotic texts, no one knows.

After Colonel Hime published his findings, some scholars claimed that the colonel was seeing ciphers in his soup. Others asserted that while the Argyle cipher had indeed been utilized, Bacon had employed an anagram code—jumbling the letters—as well to further disguise the formula. In any event, if this was not a cipher, one has to believe that a man who wrote so clearly and convincingly about flying machines and submarines had suddenly, in the final pages, become confused and incoherent.

These days of gunpowder and lenses and thought experiments were probably the happiest of Bacon's life. That he had the wherewithal to spend so much on materials and the freedom to conduct his researches at Oxford was due in no small part to the prosperity that his family enjoyed under Henry III. But Henry had grown into an utterly ineffectual king whose ineptitudes were about to catch up with him. England would once more begin the descent into civil war, causing the Bacon family's wealth to be destroyed and precipitating radical and unfortunate changes in Bacon's life.

HENRY HAD INHERITED NEITHER JOHN'S VICIOUSNESS nor his cunning, and he bungled through a reign that would last over fifty years. An indecisive ruler and poor administrator, Henry had little taste for court intrigue. A pious man interested in art and architecture, his most noteworthy achievement in his half century as king was the decision to rebuild Westminster Abbey. But grand architecture costs money, as does the upkeep of a court, and without conquest, the only way to raise this money was through taxes. Just as Bacon was refining his science, Henry's barons were growing increasingly restive.

It was one of these barons, Henry's brother-in-law, Simon de Montfort, who was to prove the king's nemesis. The son of the Simon de Montfort who had led the French crusade against the Albigensians, this Simon was brave, handsome, and intelligent, a warrior-crusader like his father. He was highly educated and on intimate terms with both Grosseteste and Adam Marsh, and he used the Oxford faculty as a sort of political think tank. He was also an extremely capable leader—much more so than Henry—and both of them knew it. Simon had once told the king that "he deserved to be shut up like Charles the Simple." By the early 1250s, Simon had begun to think that he would do a far better job of running the country than Henry. Many of the other barons agreed with him, but overthrowing a king was an extreme act. That all changed when Henry committed his greatest blunder of all. He decided to have his second son, Edmund, receive the crown of the suddenly vacant kingdom of Sicily.

FREDERICK II'S FINAL BATTLE WITH THE CHURCH had been ugly and futile. Although he had succeeded in chasing Innocent IV out of Rome and into France, the pope had initiated several assassination attempts. While able to thwart them all, Frederick, now in his fifties, had descended into paranoia and tyranny. He knew that Innocent was buying informants and allies, and fomenting rebellion all over Italy. "Anyone who showed letters from the Pope lost hands and feet," reports Ernst Kantorowicz. "The Emperor recognised rebels only, not enemies; hence every non-imperialist found armed was hanged." The repression only encouraged defections; people grew afraid of Frederick's moods and turned to the pope as the more stable alternative.

In December 1250, Frederick caught a fever while hunting at Foggia. It turned to dysentery, and in three days he was dead. The world had lost one of its great leaders, both intellectual and political. In the end, he did not change the world as he had wanted to, but he had come very, very close.

To ensure the Holy Roman Empire's irrevocable dissolution, the pope immediately offered the kingdom of Sicily to the highest bidder. There was not a great deal of interest. In addition to paying the purchase price, the new king would also have to fight off Frederick's two sons, Conrad and Manfred, who were not likely to look favorably on an attempt to snatch their birthright.

Eventually the pope approached Henry and suggested that Sicily would be a perfect realm for his young son Edmund. Although he had no money to pay for the acquisition and was ill equipped to mount a military operation thousands of miles from home, Henry, without consulting his barons, snapped up Innocent's offer. To fund the acquisition, he decided simply to, once more, raise taxes.

Just as these new taxes were levied, Roger Bacon seems to have run out of money. He did not approach his family, or if he did, funds were not forthcoming. Bacon was now firmly associated with Oxford, and Oxford was, in turn, associated with Simon de Montfort and the barons. When the barons decided to issue a manifesto of grievances against the king, it was the Oxford masters to whom they went to help draw them up. This cannot have pleased Bacon's royalist family.

But there was another alternative for Bacon to gain the wherewithal to continue his work. By the 1250s, nearly every important medieval scholar had come to be associated with one of the two orders: Albert and Thomas with the Dominicans, and Alexander of Hales, Robert Grosseteste, and Adam Marsh with the Franciscans. Despite the vows of poverty taken by each individual friar that prevented the purchase of scientific materials, each order seemed more than willing to make them available. When Grosseteste died in 1253, he left his entire library—not only his books, but all of his work, including the Greek grammar and the manuscripts on the rainbow—to the Franciscan convent at Oxford. These were now available to any member of the Friars Minor granted access by the order.

Bacon was a deeply religious man who believed in the simple moral life of St. Francis. Also, given the choice of the order of Albert and Thomas and that of Grosseteste and Adam Marsh, he was unlikely to choose the former. So, sometime around 1255, he followed his mentors, the two men he most admired in the world, renounced worldly possessions, and donned the coarse gray robe of the Friars Minor.

His timing could not have been worse. Just months after Roger Bacon put his personal and scientific future in the hands of the Franciscans, the long-simmering feud between the theology masters and the mendicants at the University of Paris exploded. To hold on to its gains at the university and control the growing schism within its ranks, the Friars Minor appointed a new leader, an autocrat called, of all things, "Bonaventura" (good luck) who believed that empirical science was a tool of the devil.

CHAPTER NINE

# Autocracy in the Order of St. Francis

. . .

BONAVENTURA WAS BORN JOHN OF FIDENZA, in Bagnoregno near Viterbo, Italy, in 1221. According to legend, as an infant he suffered from a serious illness that was cured personally by St. Francis, who exclaimed, *"O buono ventura,"* upon his recovery. He joined the Franciscans in 1238, at age seventeen, while a student in Paris studying with Alexander of Hales. He stayed on to teach, and in 1248 was granted the chair in theology that Alexander had secured for the Franciscans.

Bonaventura preached a return to the apostolic poverty espoused by St. Francis, and he lived according to those dictates himself. He revered tradition, distrusted innovation, and moved to codify Christian theology according to previously accepted theories and reasoning. He was not ignorant of Aristotle and the new learning and, in fact, quoted more frequently from Aristotle, albeit often in harsh criticism, than any other theoretician that had come before.

The crux of Bonaventura's philosophy was that all creation was a reflection of God. No part of the universe, from human beings to stars, to rocks, to falling raindrops, was to be considered in any way except as a small reflection of God's greatness. The beauty of a flower, the expanse of the heavens—the only truth to be sought from these was God's truth. All knowledge flowed through revelation and was possible only through grace. One did not reason the existence of God—one felt it. This was perhaps the closest anyone had come to St. Francis's original vision of a faith-based, anti-intellectual theology, yet Bonaventura expressed these sentiments in well-written, logical, sophisticated, and elegant prose.

Not only did this philosophy leave no room for scientific experiment, it completely negated any value of observed phenomena or material experience. And so, with the ascension of Bonaventura to minister general of the Order of the Friars Minor, the spirit of scientific inquiry that had once permeated the order, personified by Robert Grosseteste and Adam Marsh, and which had so appealed to Roger Bacon, became anathema.

Bonaventura came to power while holding the Franciscan chair of theology in Paris during a time of extreme turmoil for the orders. The cause was familiar; some students were beaten and left for dead in the streets of Paris, and the arts faculty and secular theological masters walked off the job. Once again, the friars continued to teach. But this time, remembering how much money and influence they had lost in 1229, the seculars took action against the friars. They expelled and excommunicated the three mendicant chair holders (as clerics, they had the power to do this) and proclaimed that from then on, each order could have only *one* chair, and whoever was appointed to fill it had to swear to abide by the university's rule, which meant doing what the seculars told them.

The second Dominican chair, the one the seculars had just removed, was Albert's old chair, the one that Thomas Aquinas, sitting in Paris, was waiting to fill.

The friars protested to Innocent IV, who rescinded the excommunication and ordered the seculars to readmit the three friars. The seculars refused and instead drafted a petition to be delivered to the pope by a highly respected secular theology master named William of St. Amour. William's travel expenses were paid by donations taken among all of the secular faculties.

In Rome, William dredged up every abuse by the friars that could be found—and he did not have to look far. These were not poor beggars, he argued, simple men who wanted nothing more than to preach God's word, but rather two organized and ambitious cabals populated by overfed, greedy hypocrites, perhaps even heretics, whose sole aim was to seize power and control the Church. In the apocalyptic mood of the day, William even claimed the mendicants were the forerunners of the Antichrist. (That the Antichrist had died four years before did not faze William. There was evidently always another Antichrist just over the horizon.)

William convinced Innocent, and in November 1254 the pope severely curtailed the activities of both orders. But if Innocent was with William, fortune was not. Twelve days later, before the edict could be enforced, Innocent died.

In the quickest papal election in memory, with both orders throwing their influence behind him, a new pope, the aging Alexander IV, was elected unanimously two weeks later. Alexander, who had been a cardinal protector of the Franciscans, was one of the staunchest supporters the friars had in the curia. In his second day as pope, Alexander overturned his predecessor's ruling and restored all rights to the friars in Paris.

But William enjoyed widespread support, not just among the secular theological faculty, but also among other churchmen who had grown to detest the friars' power, gained, they felt, more by intrigue than piety. The seculars ignored the pope's ruling. Both Dominicans and Franciscans were jeered at in the streets of Paris, and they feared to leave their quarters. Finally, Louis IX was forced to call out the royal archers to protect the friars.

Louis then tried to play peacemaker. He formed a committee of arbitration, and in early 1256 accepted its recommendations and declared peace. Though the terms might have tilted toward the friars, they did not represent total victory for either side. William was unmoved and, in an utterly featherbrained display of pique, launched a personal attack against the king in a sermon. He implied that Louis had been brainwashed by the friars and that he was a thief and a hypocrite besides. He advised Louis to stop allowing himself to be influenced by false messengers who should be off begging instead of trying to meddle in education or politics.

William, as a faculty member, enjoyed clerical immunity, but Louis nonetheless might have slaughtered him then and there—summary execution, even of clerics, was not an unknown phenomenon in the thirteenth century. Instead, with the forbearance that would later gain him sainthood, the king merely petitioned Pope Alexander, who was only too happy to condemn William's work and deprive him of office and benefits. Louis then banished William from Paris, Thomas Aquinas got his chair, and Bonaventura was appointed minister general of the Friars Minor.

When Bonaventura took office he was faced not simply with the threat from the seculars. Internal dissent was threatening to rip the Franciscans apart. On one side were the *Spirituales,* who were insisting on the literal observance of the original vision of St. Francis, especially the admonitions against material wealth. On the other were the *Relaxati,* who wished to temper the original rules of the order to adapt to the times. Ordinarily Bonaventura, the simplest and most pious of men, would have favored the *Spirituales*—as did Bacon—but complicating matters was the extremely mystical turn taken by many of them who had subscribed to the apocalyptic visions of Joachim of Flora.

If he was going to preserve the order, Bonaventura knew he must act quickly and decisively. He moved to brand some of the Joachimite *Spirituales* as heretics at an ecclesiastical tribunal, causing two to be condemned to prison for life. He also acted against the *Relaxati* by advancing a program to regularize rules of the order. What Bonaventura proposed was to erect "an honorable fence [to] surround the mouth and other senses and acts, deeds and morals" of the brothers. It was this "honorable fence" that was to imprison Roger Bacon.

Bonaventura's program, dubbed the "Constitutions of Narbonne," was adopted three years later, in 1260. The Order of the Friars Minor, once led merely by force of example by St. Francis, was now to be run according to strict and inflexible rules emanating from the minister general. There were new rules for begging—from whom and under what circumstances—to counter the accusation that the mendicants were simply robbers of the poor. Friars were now responsible for ensuring that money and goods from the rich would end up in the hands of the needy, not in their own pockets. (Bonaventura was not completely successful here.) The *Relaxati* were not shut out under the new rules, however. The new minister general approved of the order holding property as long as ownership was disguised and administered by a papal trustee.

For all the rules of public conduct, the cornerstone of enforcement of the new regime was, not surprisingly, censorship.

Contact with outsiders—and that included other church officials—was now forbidden without prior review from Bonaventura himself. The order was now to speak with one voice, that of the minister general. "Let no brother go to the Court of the Lord Pope or send a brother without permission of the Minister-General. Let them, if they have gone otherwise, to be expelled from the Curia by the procurators of the Order."

There were equally restrictive prohibitions on writing.

> *We prohibit any new writing from being published outside the Order, unless it shall have first been examined carefully by the Minister-General . . . anyone who contravenes this shall be kept for three days on bread and water and lose his writing . . . Let no brother write books, or cause them to be written for sale, and let the Provisional Minister not dare to have or keep any books without the license of the Minister-General, or let any brothers have or keep them without the permission of the Provincial Ministers.*

Roger Bacon was suddenly cut off, not simply from communicating his ideas to the outside world, but even from furthering them within the order. There was nothing in the forty-year history of the Friars Minor that would have allowed him to anticipate this cataclysmic turn. Unlike the Dominican order, which had been founded specifically to perform an enforcement function for the pope, and where top-down organization was vital to that purpose, St. Francis had founded an order whose very essence was the personal choice to fulfill its founder's vision. The Franciscans had therefore developed as more of a confederation than an oligarchy.

The English branch, mostly as a result of the influence of Grosseteste, had evolved very differently from that of the French or the Italian. Now all that was to change.

SHORTLY AFTER BONAVENTURA PROPOSED HIS REFORMS IN 1257, Roger Bacon was transferred from Oxford to a convent in Paris where he could be more closely watched. His work, his outspoken nature, and his known sympathy for the *Spirituales* all made him a threat to the new minister general. Adam Marsh had recently died, and Bacon was denied the protection that Aquinas enjoyed with Albert.

In Paris, Bacon was not allowed to work or study but was instead forced to perform a series of exhausting menial tasks. Although he was apparently permitted to instruct a small group of students without pay, he was denied access to the university or his former colleagues. The order did all it could to shut him off from the outside world. He would write in 1267 that he had "for ten years been exiled from [his] former University fame," and also, "for my superiors and brothers, disciplining me with hunger, kept me under close guard and would not permit anyone to come to me, fearing that my writings would be divulged to others [rather] than to the chief pontiff and themselves." This treatment—the enforced fasting, the poverty, the menial labor, the begging—took its toll on a man who was already in his mid-forties. They treated him with "unspeakable violence," he wrote, and as a result his health broke down and he had the energy to neither teach nor write.

Over those next few years, into the early 1260s, while the Franciscan Bacon languished in a small cell in Paris, the Dominican Thomas Aquinas, now one of the brightest stars in the Church, returned to Italy.

CHAPTER TEN

# Theology Becomes a Science: The Logic of Thomas Aquinas

. . .

AQUINAS HAD LEFT PARIS IN 1259 (giving up his chair to the next Dominican) and went to the priory at Naples, where he had been originally inducted into the order. As the first Italian to earn a chair in theology in Paris (just as Albert had been the first German), he was, according to an early biographer, "the splendor of the Roman province." He was made a preacher general, which gave him a voice in making policy, and was also allowed a good deal of time to study and work. Almost from the moment he arrived, he was assigned a personal secretary, Reginald of Piperno. Reginald would spend the rest of his life following his master around from place to place, taking notes or dictation and transcribing Aquinas's notoriously difficult-to-read handwriting.

Since the time of Saint Dominic, the Friars Preachers had sent a representative to the pope, a liaison between the papacy and the order, a position that had evolved into a high-level advisory post on matters of theology. Every pope was required to field questions from all over Europe on Church doctrine, and whoever answered those questions in effect set religious policy. In 1261, the order sent

Aquinas to the papal court at Orvieto to become the pope's theologian.

Intellectually, there could not have been a better man for the job. Unlike his contemporaries who merely read philosophy and applied it—most in something of a slapdash manner—Aquinas *was* a philosopher, the greatest mind in the Church since Augustine.

When he assumed his position at the pope's side, theological policy was very much up for grabs. On one side was Bonaventura, the last great Augustinian philosopher, who rejected Aristotle's logic and fought to retain the mystical interpretations that had guided Church doctrine for eight centuries. Although opposition to Aristotle was centered among the Franciscans, fear of the Philosopher was widespread, even among Dominicans. At the other extreme were those who unconditionally embraced Aristotle, or at least the Aristotle of Averroës. Proponents of this view were centered among the arts masters at Paris who were beginning to form themselves into a cohesive movement. These scholars believed not only that reason just might hold primacy over faith, but, even more threatening, that scripture needed a complete overhaul or at least a reinterpretation to bring it into line with the new science.

Aquinas could now use his position to attempt to negotiate a path through these two seemingly irreconcilable arguments. It was his genius that he understood—as probably did Bacon as well—that the war over theory was already over. It was no longer a question of whether Bonaventura and the Augustinians could hold back the Aristotelian tide, but only in what manner Aristotle was going to be incorporated into dogma.

During his tenure at Orvieto, Aquinas set himself to obtaining the fullest and most comprehensive translations of Aristotle possible. Not satisfied with the Latin versions rendered in Spain and Sicily, he began to utilize more recent translations, these from William of Moerbeke, a Dominican who would later become bishop of Corinth. It is likely that some of William's later translations were the direct result of entreaties by Aquinas perhaps even at the behest of the pope, but, in any event, Aquinas gobbled them up as quickly as William could turn them out.

From there, Aquinas began a line-by-line analysis of Aristotle's work, an undertaking not attempted since Averroës. (Albert had merely translated and explained, sometimes creatively, but had not tried to interpret.) To this Aquinas added an equally scrupulous analysis of the Arabic commentaries. This work would occupy him for the rest of his life and, when he was done, he had remade not only Aristotle but the Christian faith as well.

In one of the most brilliant leaps that any theologian would ever attempt, Aquinas decided that if Christian revelation could not be made to conform to Aristotelian logic, then somehow Aristotelian logic would have to be made to conform to Christian revelation. He met the challenge of scientific inquiry by making theology a theoretical science in itself. He produced scholastic, Aristotelian proofs for every key scriptural assertion, everything from the existence of God to the primacy of revelation, to the pseudo-Christianity of Aristotle. Although the Church did not yet know or appreciate it, Thomas Aquinas almost single-handedly gave Catholicism the tools with which to repel the onrush of empirical science. Because of him, for the next three centuries, the Church would maintain its supremacy as the ultimate authority over matters not only of heaven but of earth as well. So far ahead of any other theological thinker was Aquinas that his time is often referred to as "The Age of Scholasticism," almost as if the previous seven hundred years had not existed.

Aquinas was as prolific as he was brilliant. He produced more than sixty works, thousands upon thousands of pages. In one work alone, his famed *Summa Theologica*—unfinished when he died—Thomas posited 512 questions with 2,669 articles, and then produced proofs to dispense with more than 10,000 objections.

His organization of a problem and use of logic were dazzling—Aristotle could have asked no greater practitioner of his method. Aquinas posed a question, postulated the range of potential arguments, then systematically and painstakingly eliminated every invalid conclusion, false hypothesis, logical inconsistency, or misapplied premise until he had reduced the possibilities to just one—that which he had sought to prove in the first place. It was as if, in the densest of forests, reaching one fork in the trail after another, although he had a map (indeed, he had drawn it), he deliberately proceeded down every false path until each was exhausted, all to prove that only one correct way through the thicket existed. When trying to follow someone engaged in such an exercise, it is best not to fall too far behind, lest one risk becoming hopelessly lost. The temptation, therefore, to simply agree to certain premises in the rush to keep up, to take the guide's word for it, can be powerful indeed.

Aquinas's favorite technique for incorporating Aristotle into Catholicism was to prove that the arguments of the Philosopher and the tenets of faith were not inconsistent after all. He dealt with these issues throughout his career. A wonderful example of his method is in an *opusculum* (short work) called *De Aeternitate Mundi (On the Eternity of the World)*.

Aristotle had very clearly stated in *On the Heavens* that "the universe as a whole neither came into being nor can it be destroyed . . . On the contrary, it is unique and everlasting. It has neither a beginning nor an end . . . and it also contains within itself time without end." Finding a way to make *that* conform to Christian dogma was going to be some trick. Both Bacon and Albert had tried and failed.

Aquinas began, "Let us assume, in accordance with the Catholic faith, that the world had a beginning in time. The question still arises whether the world could have always existed . . ." Even Aquinas knew better than to try to prove logically that two contrary statements were both true, so he came at the problem from a different angle. Christian dogma said that the world was created at a finite point in time by an act of will by God, so this statement *must* be true, and therefore Aristotle's contention that the world was eternal had to be false. But what if it was false by coincidence rather than philosophic error? In other words, what if God could have made the world eternal had He chosen, but instead *chose* to create it specifically in the manner delineated by Christian faith? That would preserve the Aristotelian structure while at the same time reaffirming revelation.

And, in a clear and cogent, step-by-step process, this is precisely what Aquinas proceeded to prove.

First, he stipulated that everyone, Christians and philosophers alike, agreed that something not made by God could not have always existed. "Everything that in any way exists, cannot exist unless it is caused by him who supremely and most truly has existence," he observed. "However, someone may hold that there has always existed something that, nevertheless, has been wholly caused by God, and thus we ought to determine whether this position is tenable."

If, Aquinas went on, it is impossible for something caused by God to have always existed, it is either because God lacked the power to cause it—obviously untrue—or that something that has always existed for some reason could not be made, even though God *could* have made it. If, in turn, something God had the power to make *couldn't* be made, it was only because either it lacked the potential to be made—passive potentiality, as Aquinas put it—as steel cannot be made without the presence of iron ore or a two-by-four cannot be made without a pine tree, or because it was contradictory.

To demonstrate that the first alternative was false, that something could be made even without "passive potentiality," Aquinas used the example of angels. Since God most certainly made angels even though "no passive potentiality precedes its being," then it followed that "a passive potentiality need not precede in time whatever God may make."

That leaves the second case, which is that God cannot make that which is self-contradictory, "just as an affirmation and a denial cannot be made simultaneously true," or, as Thomist scholar Ralph McInerny phrased it, something cannot at the same time be "p" and "not p." (It is interesting that neither Aquinas nor those who study him ever accept the possibility that the laws of logic do not apply to God at all, and thus the entire structure upon which scholasticism is built is unstable.) Aquinas granted that, although some great men have "piously maintained" that God can cause contradictory events, he believed that they were mistaken. In either case, Aquinas concluded that there was no contradiction between being made by God and having always existed. "It would clearly be derogatory to divine omnipotence . . . to say that we creatures can conceive of something that God is unable to make."

Aquinas concluded *De Aeternitate Mundi* having proved the perilous, antiscriptural conclusion that God may well have made something that has always existed. At first blush, this is exactly what the Augustinians feared, a demonstration that Aristotle might be right and the Bible wrong. This, of course, was not at all what Aquinas had in mind, as Dr. McInerny points out. "While [Aquinas] firmly accepts as revealed truth, as a truth of faith, that the duration of the world had a beginning, that time and the world began . . . God might have done what he in fact did not do, namely, create an eternal world." McInerny then cites the passage in *Summa Theologica* that demonstrates the genius of this conclusion. "The arguments Aristotle puts forth are not demonstrative in the strict sense, but only broadly speaking, since what they do is disprove those arguments of the ancients which attempted to show that the world has come to be in one of the ways in which this is truly impossible."

In other words, the man who stated that the universe "has neither a beginning nor an end" was actually supportive of Christian doctrine all the time.

WHAT AQUINAS DID HERE AND ELSEWHERE was what attorneys attempt to achieve in legal briefs—present an interpretation that both fits the facts and leads to the conclusion favorable to the client. While it may have been impossible to prove that Aquinas's arguments were wrong, it was equally impossible to prove that they were right. Not once in the thousands of pages that he produced did a conclusion of Aquinas's ever result in anything other than that which he had originally sought to prove.

Aquinas intended logical construct to be a science in itself, to remove the necessity—or even the possibility—of empirical verification. There were some significant practical consequences. Because Aquinas made no distinction between method and conclusion, many of Aristotle's scientific conclusions that were just plain wrong, such as the geocentric universe and the division of matter into earth, air, fire, and water, became cornerstones of natural science within the Christian religion. Since it was often hard to define where permissible scientific inquiry ended and the truth of faith began, all of these dubious truths were now off-limits and thus inviolable.

Although *Summa Theologica* is his most famous work, another treatise, *Summa Contra Gentiles* (*Summary of the Arguments Against the Disbelievers*) had a far greater impact during his lifetime.

Aquinas began the work in Paris in the late 1250s but did not complete it until 1264. It was believed to have been written, at least in part, in response to a request by a former master general of the Dominican order living in Barcelona who, "ardently desiring the conversion of infidels," asked Aquinas "to write a work . . . that would both take away the thick atmosphere of darkness [for infidels] and unfold the doctrine of true light to those willing to believe."

*Summa Contra Gentiles* was divided into four parts, the first three consisting of that which can be known by reason and the fourth devoted to that which can be known only by faith. It did not make much of a ripple in Spain—it is unclear whether it precipitated a single conversion—but it caused an immediate sensation in centers of learning across the rest of Europe. It was perhaps the most lucid, organized treatment of Church philosophy since Augustine, and the name of Thomas Aquinas was on everyone's lips. It is inevitable that word of this great achievement would have penetrated even the most cloistered Franciscan monasteries in Paris.

In *Summa Contra Gentiles*, all of Thomas's vast powers were brought to bear—the patient enunciation, relentless logic, and unblemished piety—all to prove that which could be proved and to persuade to accept as faith that which could not. But all of Aquinas's flaws were on display as well, particularly the willingness to manipulate premise or conclusion to fit a predetermined belief.

To prove the existence of God, for example, Aquinas began by paraphrasing Aristotle's theory of the Prime Mover:

> *Everything that is in motion is put and kept in motion by some other thing . . . That mover therefore either is itself in motion or not. If it is not in motion . . . we must posit something which moves other things without being itself in motion, and this we call God. But if the mover is itself in motion, then it is moved by some other mover. Either then we have to go on to infinity, or we must come to some mover which is motionless; but it is impossible to go on to infinity, therefore we must posit some motionless prime mover . . . The Philosopher also goes about in another way to show that it is impossible to proceed to infinity in the series of efficient causes, but we must come to one first cause, and this we call God.*

The vulnerability of this argument, and one that should have bedeviled the incisive logical mind of Aquinas, is that, while the Prime Mover might demonstrate the existence of *a* god, it did not necessarily prove the existence of *the* God. There is absolutely nothing in the logical implications of the Prime Mover that led to the Christian conception of God. Rather than confront the issue, Aquinas end-ran it.

> *After showing that there is a First Being, whom we call God, we must inquire into the conditions of His existence. We must use the method of negative differentiation, particularly in the consideration of the divine substance. For the divine substance, by its immensity, transcends every form that our intellect can realize; and thus we cannot apprehend it by knowing what it is, but we have some sort of knowledge of it by knowing what it is not.*

This inquiring Aquinas proceeded to do in a series of chapters entitled "That in God there is no Passive Potentiality," "That in God there is no Composition," "That God is Incorporeal," "That God is His own Essence," "That in God Existence and Essence are the same," "That in God there is no Accident," "That the Existence of God cannot be characterized by the addition of any Substantial Differentia," "That God is not in any Genus," "That God is not the Formal or Abstract Being of all things," and "That God is Universal Perfection."

Even if each of these arguments held up under logical scrutiny, in order to differentiate the scriptural God from just any god at some point Aquinas had to shift from what God was not to what God was. In a chapter entitled "How Likeness to God may be found in Creatures," he made his leap:

> *Effects disproportionate to their causes do not agree with them in name and essence. And yet some likeness must be found between such effects and their causes: for it is of the nature of an*

*agent to do something like itself. Thus also God gives to creatures all their perfections; and thereby He has with all creatures a likeness, and an unlikeness at the same time. For this point of likeness, however, it is more proper to say that the creature is like God than that God is like the creature . . . thus the creature has what belongs to God, and is rightly said to be like to God: but it cannot be said that God has what belongs to the creature.*

In other words, just as it says in Genesis, God *must* have created man in his own image, albeit an image that is an imperfect representation. When Aquinas stated, "it is of the nature of an agent to do something like itself"—a questionable premise to begin with—he once again ascribed to God dicta that should not bind the entity that he had spent all this time describing. By what reasoning should God be bound to do something like Himself? God might have created creatures with no likeness to Himself at all if He so chose. To return to his proof in *De Aeternitate Mundi*, the only things that God was powerless to do were those that were self-contradictory—"p and not p." There was nothing at all self-contradictory about God, as defined in the Aquinas proofs, creating life that was in no way like Himself.

But who could say so? To deny the conclusion, or even to raise an eyebrow as to the logical progression if it called the conclusion into question, was heresy. The only way to attack Aquinas was within the accepted bounds of scholasticism—that is, to quibble over points of logic—and in this Aquinas was without peer. Other scholastics might parse him, Bonaventura might chide him for his affinity for pagan philosophers, and some of his anti-Augustinian conclusions might even be condemned—as they would be in 1277—but the fact remained that unless a critic was willing to challenge Aquinas's acceptance and manipulation of scriptural premises, his work was unassailable.

Roger Bacon would have one opportunity and one opportunity only to refute Aquinas and make the case for the pursuit of science, and it would come, fittingly, from the legacy of Frederick II.

SOON AFTER FREDERICK'S DEATH, his son Conrad had indeed raised an army and swept into Italy. The pope had immediately turned to Henry III to demand that he honor his pledge to send both money and troops to defend Sicily. Before Henry had a chance to try to beg off, Conrad, unused to the warm weather, died of fever in 1254.

That did not end the Hohenstaufen threat, however, as the Antichrist's illegitimate son Manfred, a far more serious contender, took up the fight. Manfred, whose mother was probably an Arab, was brave, intelligent, and charismatic. He was also a poet, warrior, and scholar, and enjoyed a fierce loyalty among his followers. He made significant gains in battle, moving up the Italian peninsula toward Rome.

The pope then tried to force Henry to live up to his deal by threatening him with excommunication and interdict. When Henry tried to raise the money and the army from his barons, they rebelled and in 1259 set up a council to run the country. The council could not get past its own petty bickering, however, and Simon de Montfort began to seize power, moving England toward civil war.

In 1261, Manfred remained unchecked and a new pope, Urban IV, son of a French cobbler, gave up on England entirely and turned to the much more powerful France. He opened negotia-

tions with Louis IX, suggesting that perhaps Louis's brother, the ferocious Charles of Anjou, might be the perfect king for Sicily. Charles had already turned down the honor once, so to help Louis in his deliberations Urban promoted a number of French clerics to high positions in the Church, especially those who had been in favor with the king.

So it came about that an obscure knight-turned-lawyer-turned-cardinal named Guy de Foulques was appointed papal legate to England, an act that was to change the course of both Roger Bacon's life and scientific history.

CHAPTER ELEVEN

# The Great Work

. . .

ORIGINALLY KNOWN AS GUY LE GROS ("Fat Guy"), Guy de Foulques was the son of a knight who had entered a monastery after the death of his wife. He grew to manhood, married, and, at first, like his father, chose the warrior life. Soon, however, Foulques discovered that he would rather persuade than slaughter and undertook the study of law. He became a noted advocate in the court of Louis IX, was appointed to the king's cabinet, and was soon a trusted councilor. When his wife died, Foulques, again like his father, abandoned the secular life and entered the Church, although not through one of the mendicant orders.

He rose in the Church as quickly as he had under Louis. Renowned for honesty, piety, and moral incorruptibility, he was appointed bishop in 1256, archbishop in 1259, and cardinal in 1261. After Urban named him papal legate, he made a number of visits to England. In late 1264, Urban once again sent him to England, where Simon de Montfort had now assumed almost dictatorial powers after defeating the hapless Henry and taking him prisoner at the Battle of Lewes the previous May.

Montfort denied passage across the Channel to Cardinal Foulques, suspecting, not without justification, that the papacy sought the return of Henry to the throne as a means of ensuring that Church authority—and property—in England would remain undisturbed. Foulques responded by excommunicating Montfort's barons and instituting a boycott on trade with England in wheat, commodities, and wine. As wine prices tripled, many English were far more sober than they wished to be, and grumbling about Simon de Montfort's government increased. Cardinal Foulques was still in northern France awaiting clearance to enter England when Urban IV died.

The year before, just as Aquinas was putting the finishing touches on *Summa Contra Gentiles,* Roger Bacon had sent a secret message from Paris, probably in cipher, to Cardinal Foulques. It was delivered through an emissary named Raymond of Laon, a member of the cardinal's staff. It is not clear how Bacon and the cardinal came to know each other, whether this message was written or verbal, or even whether it was Bacon or Foulques who initiated the correspondence. In the message, Bacon proposed forwarding to the cardinal some "writing that [he was] ready to compose, but that [was] not yet written."

Either Raymond delivered the message incorrectly or it was wrongly interpreted by Foulques, but the cardinal came away with the notion that the writing had already been put to paper. He sent a return message indicating that he was extremely interested in what Bacon wished to send him and urged him to transmit the material as soon as possible. Aware of the political delicacy of soliciting unapproved work from a member of the Friars Minor after the Constitutions of Narbonne, he charged Bacon to communicate in secrecy.

It is easy to imagine Bacon's frenzied reaction. After decades of work, he now had an opportunity to have his views heard by a man close to the pope but no immediate means to take advantage of it. It took money to produce a manuscript, and Bacon was now an impoverished friar. He could not, of course, openly petition his order for writing materials and the services of scribes to copy texts. If he wanted to obtain the wherewithal, he would have to do so privately, and so, to that end, he wrote to his brother in England, asking for funds. But his brother's fortunes had been wiped out in the civil war and he could not help.

Family assistance out of the question, Bacon proceeded to quietly scrounge about among his acquaintances, borrowing wherever he could. He seems to have done so a little at a time, eventually working his way up to the sixty pounds that it would take to finish the task. During this time, he was doubtless working on his treatise, but just how much of it was completed in the next year or two is unclear.

Then, on February 5, 1265, an unthinkable stroke of good fortune came Bacon's way. In the most unlikely of selections, to succeed Urban an astonished Cardinal Foulques was elected pope and was crowned on February 22, taking the name Clement IV.

THE NEW POPE WAS UNIQUE in having neither a theological degree nor an ecclesiastic pedigree. Cardinal Foulques was chosen because he was French and had a history in Louis's court, and Louis was now the key player in Rome's efforts to repel the Hohenstaufens. After his elevation, the new pope immediately, if somewhat reluctantly, concluded the alliance with Charles of Anjou, whom he needed politically but distrusted, even detested, personally. Charles was crowned the king of Naples and Sicily in a ceremony in Rome by a delegation of cardinals, but not Clement, who remained in Viterbo. (Clement IV was never even to set foot in Rome.)

The new king took his army to meet Manfred at Benevento, two hundred miles south of Rome. Manfred's Saracen archers gained an early advantage, but Charles's cavalry eventually turned the tide. When all was lost, Manfred, insisting on dying with honor, rode into the thick of the battle and was cut down. His own soldiers wept at the sight of his body, and afterward the French soldiers laid stones upon his grave, creating an enormous monument on the spot. After Benevento, only Frederick's twelve-year-old grandson, Conradin, remained, living with his mother in Germany.

At about the same time, the English monarchy was restored when Henry's son, the future warrior king Edward I, "Longshanks," rescued his father, defeating—and beheading—Simon de Montfort at the Battle of Evesham. In early 1266, the reinstated Henry sent an emissary to Viterbo to reaffirm his loyalty to the pope. When the emissary, Sir William Bonecor, passed through Paris on the way to Italy, he stopped to visit Roger Bacon. While there is no way to know for sure, it is possible that Bonecor was acting on Clement's instructions and that the meeting was either clandestine or disguised as a courtesy call to an old Oxford master. In any event, Bacon gave Bonecor a new letter for the pope, in which he explained that the writings he had promised two years before were not yet ready, not because of Bonaventura's prohibition but because he had had trouble raising the money.

Bacon's detractors have painted him as a solitary crank, ensconced in Paris and having little effect on anyone save the few students that he taught privately. If this was so, it is hard to imagine a king's—or pope's—envoy going out of his way to meet a mere friar and then carrying a secret message to the pope on his behalf.

Clement replied to Bacon in a letter dated June 1266, once again expressing a high degree of interest in what he had to say. He must even have had some idea of the subject, since one of the things that Clement stressed was that he was anxious to see Bacon's proposals to correct the evils in the Church. To add even more urgency, Clement, now the highest official in Christendom, restated his directive that Bacon transmit his writing in secret, and to do so regardless of any rule to the contrary by his order.

Bacon set furiously to work. He was under pressure not simply to produce material quickly for Clement, but also to counter the influence of Aquinas. Thomas, who had been regent master in Rome since 1265, had just been appointed as the senior scriptural authority at Viterbo, where he seemed certain to have the new pope's ear.

Bacon's ultimate plan was to produce a *Scriptum Principali*, essentially a vast encyclopedia, but that was too ambitious for the moment. Purely as a stopgap, he produced what he called the *Opus Majus* (*Major Work*). The urgency, the desperation that Bacon felt, the knowledge that he had only this one throw of the dice, is evident on his very first page, where he wrote, "I shall try and present to your Holiness . . . a plea that will win your support until my fuller and more definitive statement is completed."

The *Opus Majus*, which ran to more than eight hundred pages, is one of the most remarkable scientific documents ever written. Bacon presented a way of thinking, of approaching science, that is virtually unsurpassed in the thousand years since its creation. It was the most complete and incisive rendering of scientific method and philosophy since Aristotle, touching on almost every subject important in the Middle Ages—language, mathematics, philosophy, theology, health, natural science—all produced by a man forbidden to travel, working in a tiny cell eighteen hours a day. Because it was written not for a scholarly audience but to persuade a pope—Bacon scholar Jeremiah Hackett calls it "a grant application," but "white paper" is probably more apt—the *Opus Majus* has the further advantage for modern readers of being laid out in language that is surprisingly accessible. In addition, the tone of the writing provides insight into the character of this pious, impassioned, stubborn, and fractious man.

To stem what Bacon considered the cataclysmic shift in scholastic education toward legalism and sophistry as practiced by scholars like Aquinas, he proposed nothing less than an overhaul of the medieval curriculum. Specifically, he advocated substituting an objective, empirical curriculum, heavily weighted toward the study of languages and mathematics, for the rhetoric, pointless parsing, and formless logic currently in fashion. Most radically, Bacon emphasized the practical application of knowledge across all fields. "A thorough consideration of knowledge consists of two things," his opening line began, "perception of what is necessary to obtain it, and the method of *applying it to all matters* that they may be directed by its means in the proper way" (italics added).

To make his point, Bacon began the *Opus Majus* with an attack on the current method of scholarship. There were four causes of error, he asserted, "obstacles to grasping truth." These were "submission to faulty or unworthy authority; influence of custom; popular prejudice; and, concealment of our own ignorance accompanied by an ostentatious display of our knowledge." Of these, Bacon wrote, the last was by far the most pernicious.

> *For no one is so learned in nature that he knows how to be certain in regard to all the truths involved in the nature and properties of a single fly, nor does he know how to give the particular reasons for its color and explain why it has so many feet, neither more nor less, nor can he give a reason for its members and properties . . . And since in comparison with what a man knows those things of which he is ignorant are infinite, and without comparison greater and better and more beautiful, he is out of his mind who extols himself in regard to his own knowledge.*

Those afflicted with this fourth error were not to be trusted, regardless of how sincere they seemed, since they themselves might be unaware of their condition. "Men blinded in the fog of these four errors do not perceive their own ignorance, but with every precaution cloak and defend it so as not to find a remedy; and, worst of all, although they are in the densest shadows of error, they think they are in the full light of truth."

Regardless of how receptive he thought Clement might be, Bacon was taking an enormous risk in beginning the work with an attack on a status quo that represented a thousand years of evolution of Christian doctrine. Such a gamble was surely motivated in part by desperation, but it would be a mistake to underestimate the passion Bacon felt for his point of view. In 1876, an English philosophy professor wrote, "It was the assertion of freedom of thought, of the claim of science to push forward to its conclusions, regardless of fancied consequences, with implicit trust in the grand law that all truth is ultimately harmonious."

Bacon's four causes of error are as valid today as they were in 1265. If the *Opus Majus* was limited to this section alone, it would retain immense value and could well serve as the beginnings of a code of scientific ethics.

AFTER SUCH AN OPENING, Bacon must have felt the need to reassure Clement both of his own piety and that what he was proposing was every bit as orthodox and within acceptable Church boundaries as anything that Albert or Thomas might assert. Section Two ("of this plea," as he phrased it) was entitled "Philosophy" but was actually an affirmation that ultimate truth was found only in scripture and that the revealed word represented the highest knowledge of all. But science, observation, and experiment were not enemies of faith, forces to be controlled lest the power of the religion be undermined. Quite the contrary. It was the very pursuit of knowledge that led inexorably to God. "The end of all true philosophy," he wrote, "is to arrive at a knowledge of the Creator through knowledge of the created world."

Preliminaries thus dispensed with, Bacon turned to specifics. The third section, "Study of Tongues," was the first of four that specifically laid out his plan for educational reform. In this section was an entreaty that the study of Greek, Hebrew, and Arabic be added to the trivium; this was possibly the first time that a Christian scholar had advocated the formal study of comparative languages.

In Bacon's time, translation was the primary means of acquiring base knowledge, both in the sciences and in the newly resurrected philosophical works. The accuracy of any deduction or experiment proceeding from that knowledge depended largely on the quality of translation used. One of Bacon's key complaints against Aquinas and Albert, neither of whom had learned Greek, was that they proceeded from grossly flawed translations, especially of Aristotle. He considered William of Moerbeke particularly incompetent, although William's translations turned out to be a good deal better than Bacon thought. (That he commented on William at all, of course, confirms that he was aware of Aquinas's work on Aristotle.)

Bacon's advocacy of the study of language was not to ensure the accurate representation of Aristotle and the Arab commentators alone, however. Intimacy with other languages had practical applications, such as facilitating commerce across the Mediterranean. In theology, Bacon insisted that Hebrew and Greek were necessary for a full understanding of the Old and New Testaments. Lacking this, "theologians" had substituted *The Sentences* for direct study of the scriptures, and in so doing had created what Bacon considered a false theology. *The Sentences* would not provide "a thirtieth" of the knowledge as could be had through study of the scriptures themselves; yet in Paris, Bacon asserted, that was precisely the way theology was then taught.

Although Bacon included the beginnings of Greek and Hebrew grammars in the *Opus Majus,* he did not hold himself up as a supreme linguist—there is no evidence that he had any serious knowledge of Arabic, for example, nor did he claim it. He was merely asserting that the results of a deductive or experimental process could be considered accurate only if the assumptions or source material underlying that process were accurate as well, an element of scientific methodology centuries ahead of its time.

. . .

THE FOURTH SECTION, and by far the longest in the work, is entitled "Mathematics" and was, with the fifth section, "Optics," the core of Bacon's program of education. Mathematics, as Bacon used the term, was far more than simply a means of determining quantity, position, or movement through the use of arithmetic, geometry, algebra, or trigonometry. It was a philosophic base, the discipline from which all others sprang. He called mathematics "the gate and the key" to knowledge "and all things of this world" and included in this section such unlikely subheadings as "Grammar" and "Philosophy." "All categories," he wrote, "depend on a knowledge of quantity, which mathematics treats, and therefore the whole excellence of logic depends on mathematics." (This definition is not Platonic, as some have asserted. Bacon is speaking quite clearly of Aristotelian reasoning, not geometric elegance.)

It is easy to see how mathematics, when interpreted in this way, could oversee virtually all the other sciences—astronomy, anatomy, biology, geography, physics, chemistry, optics, agronomy, alchemy, and astrology. All the means of commerce, travel, growing food, and tending herds would also fall under the mathematical umbrella, as would the arts—painting, sculpture, poetry, and music. Seen in this light, mathematics was, Bacon believed, the greatest gift God ever gave to man.

But things had gone horribly wrong. "Neglect of this branch now for thirty or forty years has destroyed the whole system of study of the Latins," he told the pope. This was as close as Bacon came to an outright denunciation of Albert, Thomas, and the legalists. "What is worse," he added, invoking his fourth cause of error, "men ignorant of this do not perceive their own ignorance, and therefore do not seek a remedy."

Much of the underlying theory in Bacon's mathematics and optics sections and a good deal of the specifics came from Grosseteste's research, which was in turn derived from theories first stated by the Greeks and the Arabs. "Very illustrious men like Robert, Bishop of Lincoln," Bacon wrote, "and Friar Adam de Marisco [Adam Marsh] who by the power of mathematics have learned to explain the causes of all things . . . moreover, the sure proof of this matter is found in the writing of those men, as for example on impressions such as the rainbow, comets, generation of heat, investigation of localities on earth and other matters of which both theology and philosophy make use."

The absence of original data, however, does not take away from Bacon's analyses. It was the ways in which Bacon proposed to use the data that was groundbreaking. He was not interested in theory for its own sake—that was the crime of which he was accusing the Dominicans. Bacon wanted science *applied*. He consolidated and synthesized available material more clearly and completely than had ever been done previously, *in order to use it*. The *Opus Majus* was filled with practical applications of mathematical science, some, such as the reform of the calendar, based on sound footing, others, such as the means to ward off old age, easy to discredit (although even this was based in a homeopathic approach not all that different from health food applications of the present day).

In optics, for example, he noted the possible usefulness of lenses "to old people and people with weak eyes, for they can see any letter however small if magnified enough." Along with the written material, he sent Clement a lens to allow the pope to conduct some small experiments of his own. About twenty years after

Bacon sent the *Opus Majus* to Viterbo, eyeglasses came into use in Italy.

Bacon cast his mathematical net everywhere, sometimes with great effect and insight, nowhere more than in his section on geography. Although wider travel and increased commerce had caused something of a renaissance in geography by the thirteenth century, the most important work ever written on the subject, Ptolemy's *Geography,* would remain lost to European scholars for another two centuries. Ptolemy's other great work, the *Almagest,* on astronomy, was available, at least in part. From this and the work of Aristotle and others, Bacon derived a brilliant overall construct.

He postulated a spherical earth through which, for geometric orientation, he drew three mutually perpendicular lines meeting at the center, thus creating x, y, and z axes. For geographic division, he divided the sphere into quarters with two circumferences, one around what later became the equator, and another through the poles. He then assumed that every place on earth was the apex of a cone, and used coordinates of objects in the heavens to project these apexes onto the earth, a theoretical navigator plotting by the stars. He thus became the first man since Ptolemy (although he did not know it) to advocate the use of coordinates to identify cities, rivers, mountains, and boundaries.

Bacon went further, producing a large map on which were plotted the coordinates of many of the cities of Europe. He used Toledo as a base and then created a grid to show relative distance and location of much of the known world. That does not mean he was always right. As to the unknown world, Bacon ascribed to Aristotle's erroneous assertion that "the sea between the west of Spain and the eastern edge of India is of no great extent."

The original of Bacon's map does not survive, but a plagiarized version may have changed the course of history. Bacon's geographic theories found their way, almost verbatim—and without attribution—into a treatise entitled *Imago Mundi* by one Cardinal Pierre D'Ailly. D'Ailly's work, which was written in the early 1400s but not published until the 1480s, contained a large map exactly like the one Bacon had described in the *Opus Majus*. Sometime in the late 1480s, the *Imago Mundi* was read with great interest by an obscure Italian navigator named Christopher Columbus, who made almost nine hundred handwritten notations in the margins. The great nineteenth-century geographer Alexander von Humboldt believed that Bacon's passages about the size of the Atlantic Ocean were key to Columbus's undertaking his journey west. (There are those who claim that Columbus might not have read the work until 1494. If so, it seems odd that Columbus would take such interest in a theoretical document the veracity of which he had already tested.)

The question of whether or not he influenced Columbus aside, Bacon's use of mathematics to overlay geography revolutionized the science. He observed that at some places along a line in Egypt, no shadow was cast, while to one side of that line, shadows were cast northward, to the other side, southward. He further noted that in some of these locations, no shadow was cast twice a year (Tropics of Cancer and Capricorn). In discussing that part of the earth that is habitable, Bacon broke from Ptolemy. "I therefore insist that, though the habitable world known to Ptolemy and his followers is squeezed into a quarter of the total, far more than a quarter is, in fact, fit for habitation."

Many of these regions, Bacon went on, were not only *habitable* but *habitated*. "There is a boundless advantage in a knowledge of

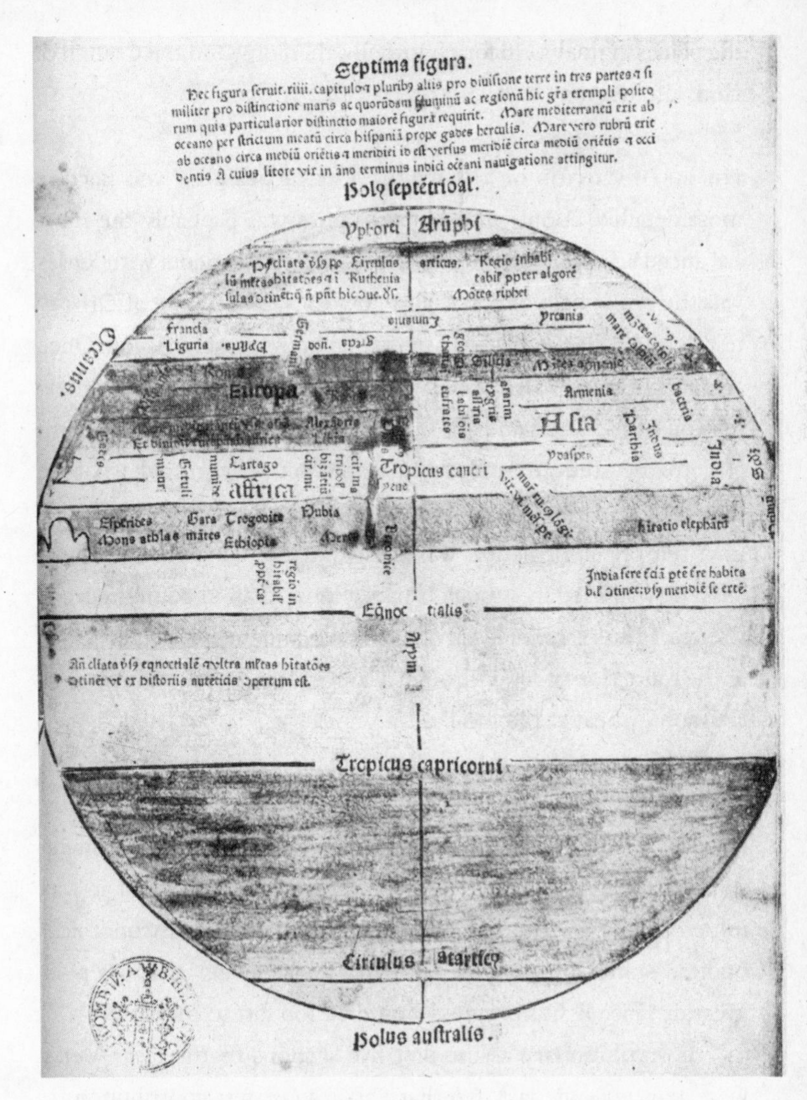


*Map from* **Imago Mundi** BEINECKE RARE BOOK AND MANUSCRIPT LIBRARY, YALE UNIVERSITY

the places in the world for philosophy, theology, and the Church of God," he wrote.

THE FIFTH SECTION OF THE WORK, "Optical Science," was Bacon's most detailed. Along with botany, optics was probably the most advanced science in the Middle Ages, and experiments with lenses was the prime field of application. Optics had been one of Grosseteste's specialties, and he had passed on his interest to Bacon. Once again, however, Grosseteste's work was largely theoretical, an extension of al-Hazen and al-Kindi. Bacon rigorously applied geometry to the study of reflection, refraction, vision, and what he called the "multiplication of forces," which was largely a theory of how sensory information was transmitted to and from humans. "Force" was used in a much broader sense than in contemporary science. It was essentially any emanation, and included such things as the refraction of light through a convex lens and the consequent creation of heat at the focal point.

In his discussion of light, Bacon made a leap, the significance of which he may well have been unaware. Departing from Grosseteste, who believed that light traveled instantaneously, Bacon asserted that light moved at a speed, but one so swift that it was imperceptible to humans. He didn't apply this to astronomical bodies—stars, for example—apparently believing that even at great distance light's progress would be too fast to measure.

As revolutionary as the first five sections of the *Opus Majus* were, it was in the last two that Bacon's greatest contribution to scientific history appeared.

CHAPTER TWELVE

# Seeing the Future: The Scientia Experimentalis of Roger Bacon

. . .

THE HEART OF THE *OPUS MAJUS*, the section of the work that most lifted it above that of Bacon's contemporaries, was Part Six, "Experimental Science." To Bacon, experimentation was a distinct discipline, separate from but vital to all others, because without experiment one could never be sure of the truth. "Without experience, it is impossible to know anything completely," he wrote. For perhaps his greatest leap of insight, Bacon employed this analogy:

> *For there are two modes of acquiring knowledge, namely, by reasoning and experience. Reasoning draws a conclusion, but does not make the conclusion certain, nor does it remove doubt so that the mind may rest on the intuition of truth, unless the mind discovers it by the path of experience . . . For if a man who has never seen fire should prove by adequate reasoning that fire burns and injures things and destroys them, his mind would not be satisfied thereby, nor would he avoid fire, until he placed his hand or some combustible substance in the fire, so that he*

> *might prove by experience that which reasoning taught. But when he has had actual experience of combustion his mind is made certain and rests in the full light of truth.* Therefore reasoning does not suffice, but experience does. *(Emphasis added.)*

This was the final break from classical scholasticism, and particularly the brand of pseudoscience that Bacon considered to have been so recently perpetrated by Aquinas. It was not enough to reason one's way to truth, no matter how sophisticated the argument. Reason followed experiment, not the other way around. "Hence in the first place there should be readiness to believe, until in the second place experiment follows, so that in the third reasoning may function," he wrote. This is the first clear statement in Christian Europe of what the modern world recognizes as hypothesis-experiment-conclusion.

To demonstrate how experimental science worked, Bacon enunciated what was probably the first rigorous, step-by-step description of scientific method ever put on paper. He did this by use of an example, laying out a series of experiments to discover the exact nature of a rainbow.

The rainbow had enormous significance in the Middle Ages. Rainbows, like thunder, lightning, and other natural phenomena, had been objects of fascination since Neolithic times. They were mentioned specifically in scripture as the reminder of a promise that God made to Noah after the flood never to repeat such a cataclysm. Genesis 9:8–16 reads:

> *And God said, "This is the sign of the covenant I am making between me and you and every living creature with you, a*

*covenant for all generations to come: I have set my rainbow in the clouds, and it will be the sign of the covenant between me and the earth. Whenever I bring clouds over the earth and the rainbow appears in the clouds, I will remember my covenant between me and you and all living creatures of every kind. Never again will the waters become a flood to destroy all life. Whenever the rainbow appears in the clouds, I will see it and remember the everlasting covenant between God and all living creatures of every kind on the earth."*

It was thus generally believed that only God could make a rainbow. It was permissible to try to deduce the nature of the phenomenon, but only with the proviso that it was divinely created. That the rainbow effect manifested itself in a variety of different circumstances—water dripping off a raised oar, light passing through a hexagonal crystal, or drops of dew on grass in the morning—only served to increase the wonder of God's presence. As a result, uncovering the nature and causes of the rainbow had become one of the preeminent scientific problems of the day, akin to modern astrophysicists trying to understand the makeup of black holes.

Bacon began with a history of rainbow theory. Aristotle, who had not been limited by the assumption of divine cause, had described a rainbow as the base of a cone in which the sun was the apex and the axis (center line) passed through the eye of the beholder to the center of the base. He believed that light reflected off individual raindrops, with colors created by different combinations of bright and dark. Both Avicenna and Averroës had discussed rainbows as well, and by the thirteenth century there was general agreement that two substances of different densities (air and water, for example) must be involved in order to create the combinations necessary to produce a spectrum.

Grosseteste had accepted the Aristotelian conical construct but claimed that rainbows were caused by refraction, with reds appearing in places where the rays were most concentrated and blues where the concentration was less, the differences being caused mostly by cloud density. Albertus Magnus had agreed that refraction was the cause but believed that light was refracted in individual raindrops, then projected onto solid material in clouds. The variety of colors, Albert asserted, was due to the difference in the density of the cloud.

From this base, Bacon then used the method he had just laid out to produce a huge advancement on anything that had come before. He returned to Aristotle's notion of reflection, theorizing that refraction was impossible since an axis *always* existed between the center of the rainbow, the observer's eye, and the sun, regardless of any movement by the observer. Then he measured the height of a rainbow when the sun was at the horizon at 42 degrees, noted that as the sun rose in the sky, the rainbow receded, and concluded that when the sun is higher than 42 degrees, it was impossible for a rainbow to appear. Then, based on the observation that people in different locations all see a rainbow if conditions were favorable, he deduced that it was reflection off a myriad of raindrops that produced the effect, not off any individual one.

Bacon used only the most primitive instruments for his experiments, and some of his deductions were incorrect. For one thing, a rainbow is formed by a combination of reflection and refraction (two refractions, actually). His own contributions, the maximum angle of the sun to the horizon and the role of a myriad of raindrops, were advances in the theory but not, obviously, the final word.* But Bacon never pretended to have the complete and accurate explanation.

Unlike his predecessors, particularly the Dominicans against whom he was struggling, his paramount assertion was that experimentation should be an ongoing process, that the search for truth does not end when one finds a convenient explanation that fits a predetermined conclusion. His aim, as he told the pope, was not to set forth the final truth of the matter—to achieve that, he insisted, further experimentation was needed—but rather to demonstrate method and plead for its inclusion in the curriculum.

He wrote to Clement:

> *Reasoning does not attest these matters, but experiments on a large scale made with instruments and by various necessary means are required. Therefore no discussion can give an adequate explanation in these matters, for the whole subject is dependent on experiment. For this reason I do not think that in this matter I have grasped the whole truth, because I have not yet made all the experiments that are necessary, and because in this work I am proceeding by the method of persuasion and of demonstration of what is required in the study of science, and not by the method of compiling what has been written on the subject. Therefore it does not devolve on me to give at this time*

*A rainbow is actually created by a process of refraction-reflection-refraction. If sunlight enters a raindrop at an angle of 42° or less to the perpendicular of the tangent of the back of the drop (thus establishing Bacon's 42° maximum height), it is refracted on the way in, breaking into component colors corresponding to the famous acronym ROY G BIV (red, orange, yellow, green, blue, indigo, violet). It is then reflected off the back of the drop and refracted once more on the way out, increasing the color separation further. Shorter wavelengths (blues) bend the most; longer wavelengths (reds) the least. If the angle is more than 42°, the light simply passes on through. As the refracted light moves back to the observer, it separates even more, so that only one color is observed from each drop. Therefore, as Bacon assumed, it takes a huge number of raindrops to create a rainbow. The conical structure is correct as well, since without the horizon to interfere, the rainbow would form a full circle.

*an attestation possible for me, but to treat the subject in the form of a plea for the study of science.*

Use of this method would open up vast new areas of knowledge. Bacon outlined this process in the "three prerogatives of experimental science." They were 1) experimental science confirms conclusions to which other scientific methods already point; 2) it reaches results that take their place in existing sciences but are entirely new; and 3) it creates new departments of science.

Here is the starkest contrast between Thomas Aquinas and Roger Bacon. Once all the questions of the extent of Bacon's actual knowledge or contribution to science and Aquinas's fundamental motivation or the soundness of his logic are cut away, this simple difference is left: Roger Bacon wanted working hypotheses to be subjected to experiment, experience, and revision, and Thomas Aquinas insisted that arguments be accepted in the abstract and on faith—in every sense of the word.

Bacon entreated Clement to apply scientific method to the production of better instruments so that observation and experimentation might proceed at a faster pace. There were immediate practical benefits to this plan. Experiments with plants, herbs, and natural substances would yield remedies that could ward off disease and extend life, and would also yield better tools and weapons. If the forces antagonistic to Christ employed experimentation first and thus created better weapons sooner, it could spell disaster for the Church.

BACON REAFFIRMED HIS COMMITMENT TO REVELATION and faith in the final section of the *Opus Majus*, "Moral Philosophy." He saw the quest for truth in science as a deeply religious act, without which there could be no genuine triumph of God. The increased knowledge gained by experiment would only serve to prove the primacy of the scriptures and discredit those who would question God's word as revealed in the Bible. It was the legalistic approach of Aquinas, denying the truth of experiment, that was a threat to God and the Church.

Moral philosophy was therefore the highest of the sciences, that to which the proper exercise of the other sciences led. Experiment, as a prerequisite to moral philosophy, would not cause man to turn away from God and faith but rather to embrace them more fully. "[Moral philosophy] in the first place teaches us to lay down the laws of and obligations of life; in the second place it teaches that these are to be believed and approved, and that men are urged to act and live according to those laws."

If the Church had adopted this view, it would have freed Christianity to be the leader in scientific inquiry without sacrificing the faith of revelation. It would have allowed the Church to promote the search for empirical knowledge within a code of scientific ethics that would have preserved the fundamental beliefs in Christ and scripture that it held dear. While certainly, as knowledge advanced, some of the literal interpretations of scripture-as-science would have come into question (as indeed they have today), the issues could have been resolved under ecclesiastic mandate. The Church, in refusing to accept this position, did not prevent scientific advancement—although it was postponed for three centuries—it merely assured that when science did regain its momentum it would be as adversary to Christianity, not partner.

THE MANUSCRIPT BROKE OFF ABRUPTLY in Part Seven. Bacon had mentioned that he intended to include a section on civil law at the end, which would have made perfect sense as a closing argument. Perhaps he ran out of time and felt the need to dispatch the manuscript to Viterbo. Perhaps he completed the section but could not have it transcribed. What is known is that even before he dispatched the *Opus Majus* to the pope he decided that perhaps it was too long or too complex and began work on a shorter version, which he called the *Opus Minus*. Later he completed yet a third version, the *Opus Tertium,* which was probably intended as a supplement. In the *Opus Tertium,* in addition to an overview of his scientific arguments, Bacon included a good bit of biographical information and social commentary. It is from this document, which was never sent to Clement, that we get much of what we know of Bacon's life and circumstances.

What Bacon did send to the pope was not simply the *Opus Majus* and the *Opus Minus,* but also an additional work, *De Multiplicatione Specierum (On the Multiplication of Species).* Although Bacon used *species* largely synonymously with *forces,* this was a far more technical treatment of multiplication of forces than that in the *Opus Majus* and was evidently included to give Clement an example of what might actually be taught in the universities. Bacon here provided more detail on his theory of perception, awareness, and how images were transmitted not only to the eye but to the brain—or the soul—as well.

To complete the package, Bacon bundled everything into the arms of his prize student, a boy known only as John, a living example of his methods:

> *The boy present, who in the midst of great poverty and with little instruction by devoting scarcely a year to increasing his knowledge has so widened his field that all are surprised who*

*know him. For I say fearlessly that although some may know more about philosophy and languages, and many may excel him in various ways, yet there are none among the Latins who surpass him in every respect, and he is a match for all of them in some things; in some points he excels them. There is no one among the Latins but may listen with profit to this boy. No one so learned, that this boy may not be indispensable in many ways. For although he has learned all that he knows by my counsel, direction, and help, and I have taught him much by written and spoken word, nevertheless he surpasses me, old man though I am, in many ways, because he has been given better roots than I, from which he may expect flowers and wholesome fruits which I shall never attain.*

And so, in mid-1268, John set off for Viterbo.

Centuries later, in a backlash precipitated by the cult of personality that grew up steadily around Bacon's memory, and particularly by the assertion that he was a man ahead of his time, some scholars chose to deprecate the magnitude of Bacon's achievement with the *Opus Majus*. They pointed out that the manuscript was riddled with factual errors and inaccuracies and that Bacon subscribed to superstition, attributing magical powers to astrological bodies, for example. They observed that there was a paucity of original research and that much of his work was derivative, merely an extrapolation of the thinking of men like Robert Grosseteste and Peter Peregrinus. They argued that when Bacon used the terms "experiment," "experience," and "mathematics," he did not do so in the modern sense but in a narrowly medieval context. They said, moreover, that he was unduly critical of his contemporaries, particularly Albertus Magnus, who was, after all, as interested in the natural sciences as was Bacon himself. They claimed he was jealous and bitter, particularly of those scholars with advanced theological degrees from Paris, who had justifiably gained the worldwide respect and repute that Bacon himself craved, and that this colored his judgment and biased his conclusions.

The wonder of the *Opus Majus* is that it is possible to grant each of these denunciations and still be awed by the depth of Bacon's achievement. For though it is true that he built on the work of others—which he never denied and indeed went to great pains to point out—no one else drew knowledge together into as coherent and persuasive a whole. The Arabs hadn't done it, Grosseteste hadn't done it, Albert hadn't done it, and Aquinas certainly hadn't done it. What is so crucial about Bacon's work is that it created a clear signpost to the future, a method by which to gain a deeper, more significant, more profound knowledge of both the world in which we live and the nature of the eternal. When Stephen Hawking concluded *A Brief History of Time* with the words "If we find the answer to [why it is that we and the universe exist] it would be the ultimate triumph of human reason—for then we would know the mind of God," he was following the trail blazed by Roger Bacon.

BY THE FALL OF 1268, Bacon's prospects for precipitating reform in the Church never seemed better. In the spring of that year, Frederick's grandson Conradin, "the last of the Hohenstaufens," then fifteen, had come to Italy to raise an army to retake Sicily from Charles d'Anjou. In August he was defeated at the Battle of Tagliacozzo. Conradin escaped after the battle but was soon captured, and two months later, despite Clement's personal abhorrence, Charles had him publicly beheaded. Still, the papacy was secured against the invader, and Clement could finally turn his attention to internal reform.

Clement, with his keen mind, determination, and deft political skills, might have been one of the great popes, perhaps even rivaling Innocent III himself. But in November 1268, after less than four years in office, his ambitions unfulfilled, Clement died, and with him Roger Bacon's opportunity to blend scientific curiosity into theology died as well.

CHAPTER THIRTEEN

# Knowledge Suppressed: The Conservatives Respond

. . .

WHETHER CLEMENT LIVED TO SEE THE *OPUS MAJUS*, the *Opus Minus*, and *De Multiplicatione Specierum* is not known. Still, it seems that the messenger John, who afterward disappeared into history without a trace, completed the task to which he had been assigned, for Bacon's manuscripts certainly reached Viterbo.

In 1270, a Polish mathematician and member of the curia named Witelo produced a large, detailed work on optics. This *Perspectiva,* dedicated to Aquinas's friend and translator William of Moerbeke, became the most popular reference work on the subject for the next four centuries. In it, Witelo borrowed heavily from Alhazen and Grosseteste, but, most strikingly, applied geometry to light and the rainbow in a way markedly consistent with that of the *Opus Majus* and *De Multiplicatione Specierum*. "Some of the parallel passages could have come from no other source," A. C. Crombie stated flatly in his authoritative work on Grosseteste. Once again, as with the *Imago Mundi*, Bacon's work was used to push forward an important scientific discipline, although he himself received no credit.

But perhaps Bacon's writings had a greater impact on his contemporaries than most medievalists believe. Certainly Bonaventura came to know of Bacon's writings, at least when Witelo published. In fact, the reason we know that Bacon's work reached the pope is Witelo's dated notation on the copy of the *Opus Majus* that is now in the Vatican Library. Both William of Moerbeke and Aquinas were also in Viterbo and would surely have known of Bacon's writings.

Perhaps because he was no longer seen as a threat, Bacon was allowed to work openly in the aftermath of Clement's death. Now in his late fifties, still in Paris, he continued to put in eighteen-hour days, subsisting mostly on scholarship and prayer. Over the course of the next four years, he produced Latin and Greek grammars and resurrected his plan to produce an encyclopedia of philosophy. Working openly did not mean acceptance, however. The battle for the soul of Christianity was reaching its climax, and, with his patron dead, Bacon was no longer a part of it.

AFTER CLEMENT, the papacy remained vacant for three turbulent years. Into this vacuum rushed all the old conflicts. Politically, Charles of Anjou justified Clement's distrust by attacking and overrunning Italy. With no Hohenstaufen to restrain his ambitions, the power of the Church was again menaced from the outside, this time by Charles's secular monarchy.

Nonetheless, it was internal struggle, spiritual anarchy, that was the real threat. The same two distinct but overlapping lines of battle re-formed at the University of Paris. The first was a continuation of the attacks on the mendicants by the secular theology masters. The seculars pressed the same charges as they had twenty years before, but it soon became clear that the orders had grown too strong. Reduced to engaging the best of the friars in a series of disputations—including Bonaventura and Aquinas, who had been sent back to Paris specifically to blunt the threat—the secular masters were eventually forced to retract many of their charges, and their challenge collapsed.

The second battle, centered in the arts faculty, was far more significant, and once again Aristotle was at the center of the storm. In about 1265, an arts student named Siger of Brabant had obtained his degree and begun to teach. Siger and his followers began to teach Aristotelian philosophy filtered strictly through the commentaries of Averroës. These "Averroists," as they were called, openly accepted two of the three Aristotelian premises that most threatened orthodox Christian theory—that the universe was eternal, and that there is one overall soul into which all individual souls merge upon a person's death.

As for the third premise, the one that postulated one truth for reason and another for faith, the Averroists slickly sidestepped the issue by trying not to talk about it. They claimed that when faith and intellect collided, they accepted Christian doctrine, but no one on the theology side took them very seriously. It was difficult for those who had elevated Averroës to a pinnacle just below that of the Philosopher himself to convince people that they did not ascribe to the single most fundamental tenet of his interpretation. Siger himself was sometimes a bit more direct than he should have been. At one point he stated that his goal was "to discover simply what the philosophers and especially Aristotle have thought, even if by chance the thought of philosophy did not conform to revelation."

Siger recruited a large following as more and more students poured in from all over Europe. These were rough-and-tumble days in Paris. The student body was divided into "nations"—French, Picards (including Belgians like Siger), Normans, and English (which included Italians, Germans, and just about everyone else)—and it was not uncommon for students from one nation to brawl with, set upon, or even kidnap students from a rival nation. Siger was intelligent and charismatic, a leader in his group, but once he began to teach, he became extremely popular across factional lines. At one point, although still only in his twenties, he was nominated for rector. Without someone to countervail the Averroists' burgeoning popularity, they might have overrun the university, sweeping the theology faculty along in their wake.

Thomas alone understood that where the battle between the seculars and the mendicants in the theology faculty was for short-term political power, the outcome of this dispute would determine the future course of Christianity. If the Averroists succeeded, literal acceptance of scripture was over. Empirical observation, experiment, and imagination would have been set free, and ultimately Christian dogma would have been forced to accommodate science.

In one of the Dominican chairs as regent master, Thomas took it upon himself to turn back the assault. He was at the very height of his powers and influence. He threw himself into his work so feverishly that it seemed as if he wanted to fight the war on all fronts at once. He worked on as many as four different books or papers at a time, dictating to the army of secretaries that were now at his disposal. In three remarkable years, he completed commentaries on Aristotle, a number of *opuscula,* and the second and third parts of *Summa Theologica*. Unlike *Summa Contra Gentiles,* this new epic was not directed at non-Christians but was instead an attempt to provide a guidebook for young theological students so that they could better understand the rules and origins of their faith.

To counter Siger, he dashed off *De Unitate Intellectus Contra Averroistas* (*On There Being Only One Intellect, Against the Averroists*). Here, Thomas demonstrated that the seeming inconsistencies between Aristotelian philosophy and Christian faith were as a result of Aristotle's being interpreted incorrectly by Averroës. In his ignorance, Thomas implied, Siger had perpetuated these errors. But Siger was an adept logician as well. He even wrote a defense of Averroës's interpretation of Aristotle, entitled *Against Those Famous Men of Philosophy, Thomas and Albert*.

Aquinas did not get a chance to finish the debate, for at the end of 1271 he was transferred back to Naples. Thomas, for all his talents, had become a figure of some controversy himself. There were many in the Church who did not understand that it was inevitable that Aristotle and science must be somehow incorporated into dogma. They saw Thomas, with his advocacy of the Philosopher, as a danger to the Church every bit as great as Siger.

Although Thomas had failed to halt the Averroists' growing popularity completely, he did succeed in creating sufficient countermomentum to establish a stalemate. As it turned out, that was all that the Church needed, although Thomas would not live to see his methods vindicated.

In Italy, he continued to work on his various projects with the usual fervor. Then, on the morning of December 6, 1273, during Mass, Thomas had what has been described as either a mystical vision or a mild stroke. In either case, from that day forward he put down his work and refused to write or dictate another word. When pressed, he said that after what he had seen, "all that I have written seems like straw to me." Over the next months, he slipped

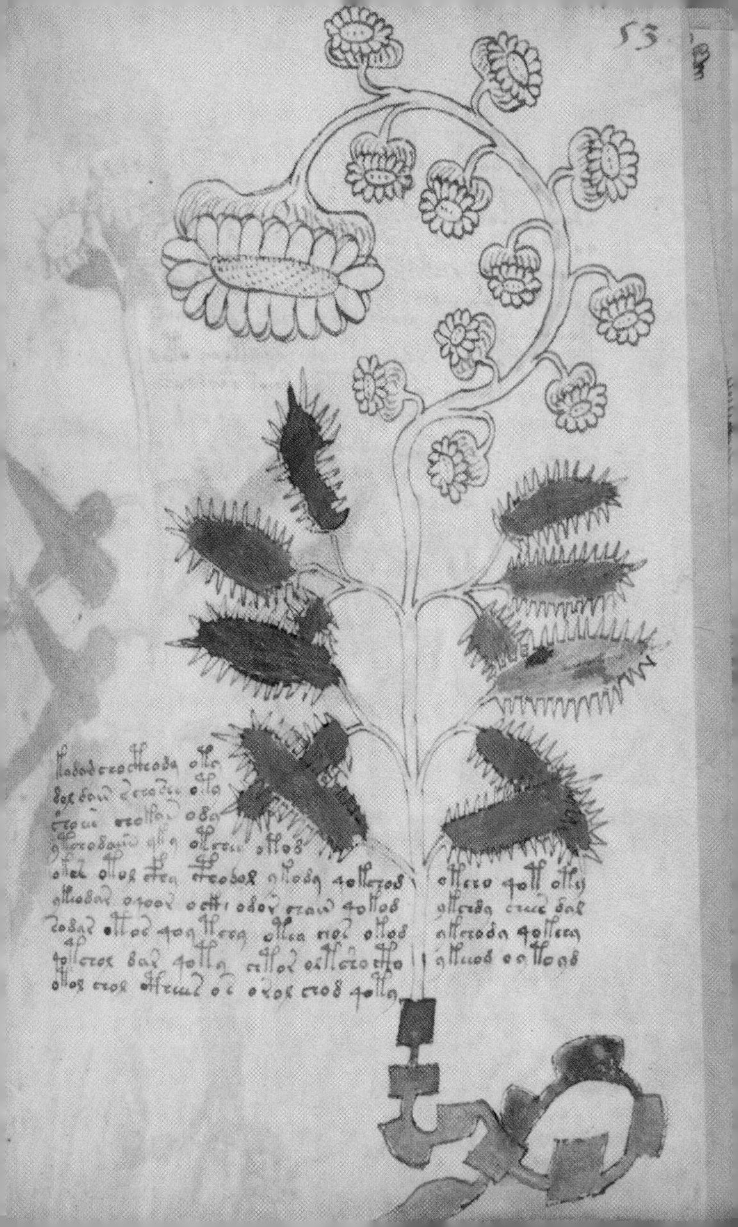
53

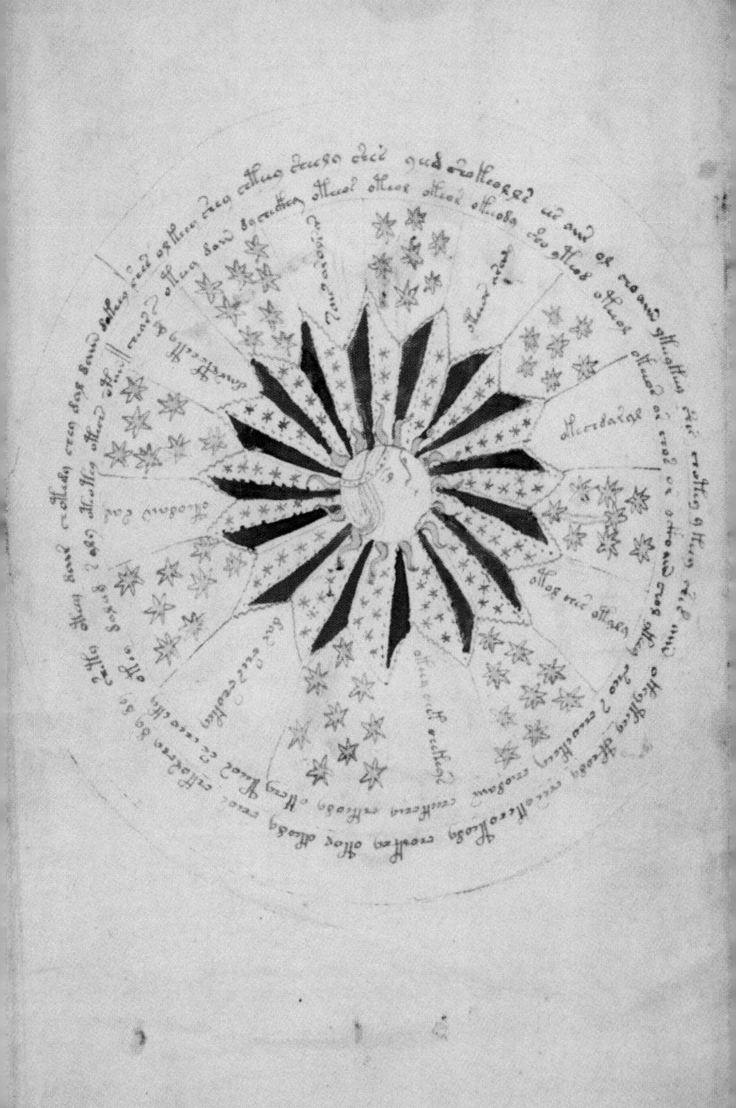

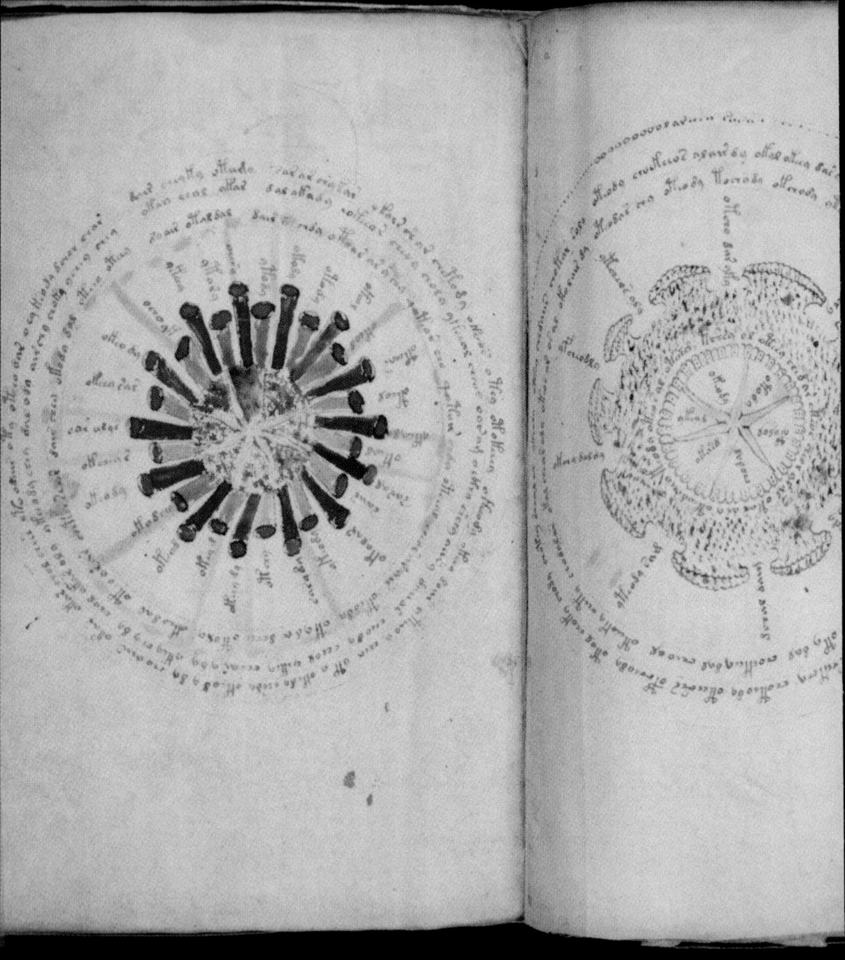

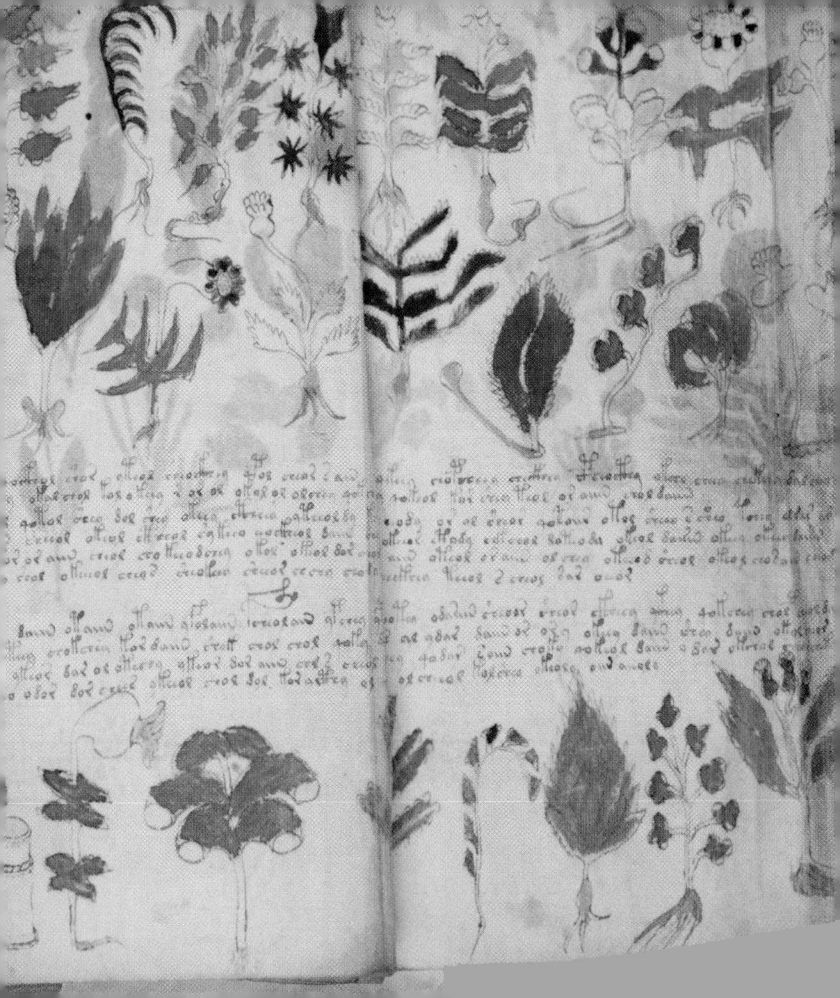

into what his acolyte Reginald described as "stupefaction." He had trouble maintaining his balance while walking, and his speech was impaired as well.

In early 1274, a new pope, Gregory X, called for an ecumenical council to be held in Lyons to try to blunt the growing territorial ambitions of Charles of Anjou, now King Charles I of Naples. Gregory had already chosen a new Holy Roman Emperor, someone who everyone agreed would make no trouble, a nondescript German knight named Rudolf of Hapsburg. (Rudolf himself proved malleable enough, but his descendants were a different story. The Hapsburgs went on to become one of the longest-lived and perhaps most acquisitive dynasty in European history, dominating at various times Germany, Spain, the Netherlands, Hungary, and the Balkan States, and surviving as a ruling family until World War I.)

The pope asked Aquinas to be present at the council. On the journey, while riding on a mule, Thomas was struck in the head by an overhanging branch. He was taken to a Cistercian monastery, where he lay ill, and then, on March 7, 1274, Thomas Aquinas died.

The council in Lyons was to claim another prominent victim. On July 15, while attending the same event, Bonaventura died as well, rumored to have been poisoned, although no suspects were ever brought forward.

With Bonaventura and Aquinas dead, and Albert now almost totally senile, Roger Bacon had lost his three greatest adversaries. He made one last attempt to have his ideas heard. He completed the *Compendia Studii Philosophii,* the first installment of his encyclopedia, and addressed it to Gregory.

The *Compendia* was similar in theme to the works that he had sent to Clement, but was written in a far less accommodating style. In addition to detailed sections on optics, language, and scriptural inaccuracy, the *Compendia Studii Philosophii* included long diatribes in which Bacon vented his fury at the mendicants, the secular masters, the arts faculty, and what he saw as a theology run amok.

With a lack of political awareness, surprising even for him, he attacked without restraint, saving his most vitriolic passages for friars and lawyers. Bacon was particularly upset that the orders continued to recruit boys in their teens, indoctrinated them with questionable dogma, then, before they had studied theology—or much of anything else—sent them out to preach. They knew nothing of scripture, nothing of language, and therefore could not help but corrupt the Church. Yet, filled with brothers with no genuine training, through political chicanery alone the orders had come to dominate the universities. It was said that Bacon maligned both orders as if he were a member of neither. It is a distinct possibility that by this time that is precisely how he felt. The Franciscans of Bonaventura certainly bore little resemblance to those of Grosseteste or Adam Marsh.

As to the lawyers, the encroachment of "civil law" into Church affairs had corrupted the religion. By "civil law," of course, Bacon did not mean jurisprudence but rather the theological legalism of Aquinas.

At a time when the Church saw itself as under assault and was seeking stability and a strategy that would both preserve tradition and accommodate radical new thinking, Bacon's blanket denunciations could only have hardened those in power against him. However, it is not difficult to sense Bacon's passion, desperation, and despair underneath the bitterness. This was an extremely pious man, forceful in his opinions, surely, but also humble in his personal ambitions. There is not a single line or passage in any of his works in which he asks for favors, preferential treatment, personal advancement, or political power.

The *Compendia* today exists only in fragment, and it is unknown whether or not it reached the pope or even was dispatched. But certainly Bacon's superiors in the order knew of it, for it was soon after the *Compendia* was completed that they decided to silence him permanently.

TO SUCCEED BONAVENTURA, the Franciscans chose a reactionary, Jerome of Ascoli. Almost immediately, some of the more prominent *Spirituales* were imprisoned, placed in solitary confinement on a diet of bread and water, and denied the sacraments. This was merely the first step in a conservative backlash against not just apocalyptic vision but also Aristotle, logic, science, and experiment. So sweeping would this movement become that not only was Roger Bacon swallowed up but even the philosophy of Thomas Aquinas would be threatened with condemnation.

The conservatives began their assault in 1277 under yet another pope, the intransigent John XXI. John, who had begun his career as a former Paris arts master of all things, knew what went on in Paris as much as anyone. He instructed the city's bishop, Stephen Tempier, to look into "reports" that certain errors of doctrine were being taught at the university. Since the university was "the living source of the wisdom of faith which . . . [carried] the Catholic faith to the ends of the world," as John put it, he asked Tempier to report to him at once if these rumors were true and, if so, who was perpetrating them.

John had chosen the right man for the job. There was no one

in Europe more anxious to sniff out heterodoxy than Stephen Tempier. Seven years before, as rector of the theological faculty, Tempier had taken it upon himself to issue a list of thirteen offensive propositions that were to be "condemned and excommunicated together with all who should knowingly teach or affirm them." All the basic Aristotelian precepts and their Arab commentaries were denounced.

At the time, to Tempier's fury, no one paid the least bit of attention. Neither Aquinas nor Siger cared to be interrupted in their quarrel, nor did either want Tempier's help. What seemed to be a silly list of thirteen propositions was brushed aside, their author was ignored, and Tempier spent the next seven years stewing over the response (or lack of it) to his efforts to clean up the university.

Now, with the pope behind him, granting him the opportunity to save the Church from the godless pagans (to say nothing of almost every Christian scholar then at the school), Tempier, by all accounts a narrow and tyrannical man, set eagerly to work. He recruited a team of like-minded zealots and proceeded to gather every example of doctrinal deviance in Paris. Thirteen general precepts became a full 219 propositions. Such was his rush to produce his list and send it off to John that he did not bother to edit the contributions of his followers, and thus certain items appear again and again in different sections.

There is little question that the members of Tempier's truth squad used the opportunity to settle doctrinal scores. Siger was not mentioned specifically, but he and his followers were prime targets. Some of the entries seemed to be directed at Bacon as well, or at least at anyone who believed that experimental science should be a part of theology. A surprising target, however, was Aquinas himself. At least sixteen Thomist propositions made the list, most concerning the Aristotelian relationship between form and matter.

Then, not waiting for John to approve the list—in fact, even before he dispatched it—Tempier publicly condemned all 219 errors and excommunicated their authors. The pope, pleased to have someone else doing the dirty work, did not reprimand Tempier for his precipitous action but instead accepted the list and used the opportunity to rid the Church (and the orders) of its most visible and charismatic opponent. Siger was summoned before the pope, condemned by the Inquisition (still dominated by the Dominicans), and held as a prisoner by the curia. Less than a year later, he was dead, stabbed by the crazed "assistant" who had been assigned to him by his enemies.

Tempier's actions were soon copied by other conservatives seeking to rid themselves of irritating opponents. Eleven days after Tempier published his list, Robert Kilwardby, the archbishop of Canterbury and a Dominican, issued a condemnation of thirty propositions of his own. At least three Thomist precepts made the English list as well. Criticism of Thomism was on the rise as conservative Churchmen—who had never really understood what Aquinas was talking about in the first place—now rushed to publish criticism of a man no longer able to respond. What they wrote was often little more than the parsing and tortured logic of which Bacon had always been so critical—the kind of arguments that Thomas, if he had still been around, could have destroyed in a breath.

Then the Dominicans rallied behind their dead star.

Toward the end of 1277, John of Vercelli, master general of the Friars Preachers, met with Jerome of Ascoli to try to smooth over the differences between the orders to present a stronger face to the common enemy. Part of the agreement was a mutual non-aggression pact in which "any friar who was found by word or deed to have offended a friar of the other Order should receive from his Provincial such punishment as ought to satisfy the offended brother." This order applied whether the offended brother was living or dead.

In 1278, at a meeting of the general chapter of the Friars Preachers in Milan, two theology masters were ordered sent to England to investigate those "who *in scandalum ordinis* showed disrespect for the writings of the venerable Friar Thomas D'Aquino." They traveled with power of excommunication or exile for anyone found guilty, meaning anyone who tried to enforce Kilwardby's edict, the archbishop himself being exempt. The next year, at a meeting of the general chapter in Paris, any criticism of either Thomas or his work was expressly forbidden. A few years later, the Paris chapter ordered every member of the order to promote Thomist teachings. Any friar failing to adhere to this edict—even those who had honest disagreements with Thomist philosophy—was to be immediately suspended. Members who demurred found themselves barred from promotion and ostracized.

Finally, at a meeting of the general chapter in 1313, it was determined that "no one should dare teach, determine, or respond differently from what is commonly thought to be [Thomas's] teaching." In addition, from that point forward, no Dominican could be sent to the University of Paris without at least three years of Thomist study.

It all worked. The momentum generated by Aquinas's brother Dominicans spread throughout the Church, and the man whose work was condemned in 1277 was made a saint, virtually without opposition, less than fifty years later. (Albertus Magnus had to wait until 1931 for a similar honor.) "By the time Thomas was canonized in 1323," according to Weisheipl, "almost all Dominicans had made the teaching of Thomas their own and considered it a privilege, as well as an obligation, to study and defend it."

The logical elegance of Thomism proved inexorable, forever changing the manner in which the Church promulgated doctrine. Over the next few decades, the work of Augustine—and Plato—slipped away in theology and even more so in science, and Aristotle reigned supreme for the next six centuries. Although Neoplatonism popped up every now and again and Plato remained a force in philosophy, it simply became axiomatic that objective classification arrived at through logical analysis led to scientific truth. There would not be a significant scientific thinker who did not use the fundamental Aristotelian method. Newton was an Aristotelian, as were Copernicus, Galileo, Kepler, Vesalius, Liebniz, Benjamin Franklin, and Charles Darwin. Even Francis Bacon, who was to pick up his namesake's mantle three centuries later, and who denounced Aristotle in the most vitriolic terms, was himself unmistakably Aristotelian.

Then, in 1905, Einstein came along and proved to a stunned and disbelieving world that it was Plato who had been right all along.

ROGER BACON, whose order was all too happy to be rid of him, received far different treatment. For him, the combination of Tempier's condemnation, the steady stream of popes who were members or sympathizers of the orders, and the Franciscan-Dominican treaty turned out to be disastrous.

It seems likely that he returned to Oxford after publishing the *Compendia Studii Philosophii*. The vast majority of Bacon's manuscripts have been found in England, which tells against his having spent the rest of his life in Paris. But Oxford could not—or would not—protect him from the repressive policies of Jerome of Ascoli.

According to a 1370 Franciscan record, *Chronicle of the Twenty-Four Generals* (of the Order of Friars Minor), this is what happened to Roger Bacon around the year 1278:

> *This Minister General brother Jerome by the advice of many friars condemned and reprobated the teaching of Friar Roger Bacon of England, master of sacred theology, as containing some suspected novelties, on account of which the same Roger was condemned to prison,—commanding all the friars that none of them should maintain this teaching but should avoid it as reprobated by the Order. On this matter he wrote also to Pope Nicholas (III) in order that by his authority that dangerous teaching might be completely suppressed.*

Although the *Chronicle of the Twenty-Four Generals* is considered suspect by some, most of the information contained in the record has proven to be reliable. What we do know is that, for the next fourteen years, the prolific writer wrote no books, the famed teacher taught no one, and the notorious church gadfly irritated not a soul.

If as is likely he did go to prison, the circumstances are unknown. Most of the *Spirituales* punished by the order during Jerome's reign were sent to a prison in Ancona, Italy, where they were kept in chains, forbidden contact with anyone including their jailers, and deprived of the sacraments, even on their deathbeds. Those who sympathized with these convicted heretics were also treated severely. According to an eyewitness, a friar who spoke out against the cruelty of those imprisoned was himself jailed and subjected to such harsh treatment that he died within months, his body discarded at the side of the road without burial.

Some scholars have hypothesized that Bacon's imprisonment would not have been so severe—that it would have been merely a form of house arrest. Even the mildest sentence, however, would have denied him outside contact and the ability to work, which, for an active, questing mind like Bacon's, would have been the worst punishment of all.

In 1288, Jerome of Ascoli was raised to the papacy, taking the name Nicholas IV. This left a vacancy at the top of the Franciscan order, and the next year, despite opposition from Nicholas, the Friars Minor elected Raymond of Gaufredi as minister general. Raymond was a much more tolerant leader, sympathetic to the *Spirituales* and their desire to maintain the simplicity and piety of St. Francis. Among his first acts of office was a review of the sentences of those condemned under Jerome. He visited Ancona, witnessed the horrors of the institution firsthand, then immediately commuted the sentences of all of the condemned friars.

It may be just coincidence, of course, but two years later Roger Bacon resurfaced back in England with his final work, *Compendium Studii Theologiae* (*Compendium of Theology*). He began by noting that until recently he had been prevented from publishing, "as is known to many."

Bacon was seventy-eight years old. This *Compendium* is a sad, confused, and tired work, written by an old man. The fire is gone from the pages. There are rambling passages that discuss his earlier work and life, but sometimes he gets his dates wrong. There are regrets about a number of experiments that he had never had the chance to make, and a passage in which he acknowledges that "the principal occupation of theologians is concerned with questions . . . and theological disputes [that] are settled by means of authorities and arguments"—in other words, that his insistence on using experiment in order to be certain has been passed by.

Bacon never finished this last work. According to a fifteenth-century Warwickshire chronicler, "The noble doctor Roger Bacon died A.D. 1292 on the Feast of Saint Barnabas the Apostle [June 11] and was buried in the Grey Friars church in Oxford." A sixteenth-century Oxford historian recorded in a royal manuscript that, after his death, all of Bacon's works that were on hand at the monastery in Oxford were nailed to the walls by the friars and left there to rot.

CHAPTER FOURTEEN

# Enigmas and Espionage: The Strange Journey of Dr. Dee

. . .

IN THE 250 YEARS SUBSEQUENT TO HIS DEATH, Roger Bacon's work lay in the moldering dank of various English monasteries. Considering the magnitude of the forces acting to eradicate his intellectual legacy—fear, ignorance, plague, scholasticism, neglect, time, decay, and, ultimately, politics—the wonder is not which manuscripts can be credited to Bacon but that any of his work survived at all. That Roger Bacon was rescued from obscurity and even experienced something of a revival can be traced to the efforts of one man, the Elizabethan John Dee, a noted mathematician, astrologer, and geographer who talked to angels.

History is full of droll characters who exist on the periphery of great events, but even in this category Dee stands out. Born in London in 1527, the son of a well-to-do textile merchant (although Dee would later claim an ancestry stretching back to an ancient Prince of Wales), Dee began his career brilliantly. His father sent him to Cambridge when he was fifteen to study Latin with the reigning English academic, Sir John Cheke. England had broken with the Catholic Church in Rome; the Protestant Reformation was sweeping over Europe. Cambridge was a Protestant center, and Cheke was one of the leaders of the movement. His list of pupils was extensive and illustrious—Cheke trained Roger Ascham, who in turn tutored Edward VI; Cheke himself tutored Queen Elizabeth. Both William Cecil, later Lord Burghley, principal secretary of state under Elizabeth, and Sir Francis Walsingham, who would play a key role in Elizabeth's administration (and this story), studied at Cambridge under Cheke. Cecil even married his sister.

From his earliest days at Cambridge, Dee developed an obsession with Roger Bacon. Despite—or perhaps because of—the efforts of conservatives within the Church to eradicate his legacy, Bacon's legend had grown over the years, but it was not the legend of a brilliant, persecuted practical scientist. The Roger Bacon of sixteenth-century folklore was a magician and mystic, the Roger Bacon of mysterious explosions in the night, talking brass heads, magic potions, burning lenses, and unbreakable ciphers.

A driven student fixated on mathematics and its applications, Dee sought the real Bacon, unearthing and then poring over whatever material had survived. He bought up every Bacon manuscript or fragment his father's fortune could afford. He sought to mold himself in Bacon's image, even copying Bacon's punishing schedule of eighteen hours a day of work and only six for meals and sleep. As Roger Bacon was interested in machines, so was Dee. For a prop in a Greek comedy he produced at Cambridge, he built a giant beetle using ropes and pulleys (for the hero to ride, instead of Pegasus). It actually seemed to fly around the stage, to the astonishment of his sixteenth-century audience.

After Cambridge, he went to the University of Louvain, where he studied trigonometry, a subject not yet taught in England, toyed with Copernican astronomy, and made the acquaintance of one of the most famous cartographers in history, Gerard Mercator. Dee and Mercator became inseparable: "It was the custom of our mutual friendship and intimacy that, during three whole years, neither of us willingly lacked the other's presence for as much as three whole days," Dee would later write. From Mercator, Dee picked up a lifelong interest in exploration, geography, and astronomy. While in Brussels, he also tutored the English ambassador Sir William Pickering (another Cheke protégé) in mathematics and the use of the astrolabe and turned down a position in Charles V's court as the emperor's personal philosopher.

From Brussels he went to Paris to lecture on Euclid, "a thing never done publicly in any University of Christendom," and had the satisfaction, at twenty-four, of causing such a sensation that students and masters hung out of the school windows in order to catch his words. "A greater wonder arose among the beholders, than of my Aristophanes Scarabeus [the mechanical flying beetle] mounting up to the top of Trinity Hall in Cambridge," he recalled later. After Paris, he returned to England, where Henry VIII's only son, Edward VI, now ruled. He got a position tutoring the sons of the Duke of Northumberland, head of Edward's Privy Council. (One of Dee's charges was the future Earl of Leicester, Elizabeth's suitor Sir Robert Dudley.) In addition, Sir John Cheke personally introduced him to his brother-in-law William Cecil, who in turn presented Dee to King Edward.

Everywhere he went, Dee bought old books and manuscripts focusing on science and mathematics—particularly by Roger Bacon, if he could find them. He had also developed a taste for the occult—even at this early date, Dee's mathematics was veering off uncomfortably toward numerology and mysticism. "By Numbers, a way is had, to the searchyng out, and understandyng of every thing hable to be known [*sic*]," he wrote in his "Mathematicall Praeface" to Euclid's *Elements*. Unfortunately, Dee's championship of mathematics had as much to do with conjuring as computation—his definition of "hable to be known" focused on the shadowy world of angels and demons. When Dee spoke of charting the heavens, he meant it literally. When he and Mercator drew up a series of Copernican models of the heavens, Dee added a couple of rings for spirits and angels.

He also believed in what was then called "judicial astrology," that the position of the stars affected life on earth. He cast horoscopes for many people, including King Edward and the Princess Elizabeth—he made something of a living at it, later on—but his philosophy went further even than this. According to Dee, everything in heaven and earth had a number assigned to it. Determine the number and you could control events: "Easy it is to be gathered, that *Number* hath a treble state: One, in the Creator: an other in every Creature (in respect of his complete constitution): and the third, in Spirituall and Angelicall Myndes, and in the Soule of man," he asserted. For these views, he was dubbed "the Great Conjuror."

In July 1553, Edward VI died, his half sister Mary succeeded him, and Dee's comfortable sinecure abruptly disintegrated. The queen who became known as "Bloody Mary" had been forced to secure her throne by successfully fighting off a Protestant scheme by Dee's old employer, the Duke of Northumberland, to crown Lady Jane Grey in her place. When the plot ultimately failed, the Protestant English nobility fled the country—a prudent idea, as Mary, in an attempt to reintroduce Catholic ritual, took to burning heretics. She managed to execute about five hundred in all during her five-year reign. Those she did not execute were thrown in the Tower or ruined.

Dee, while not specifically Protestant himself (mystical mathematics defied religious classification), was almost certainly a Protestant sympathizer. His father, Roland, was known to have been associated with the Northumberland scheme and was among the first arrested by the new queen. But Dee did not flee the country. He did not have the means to do so. Although his father was released from prison after only a month's stay, he was ruined financially, and with him, his son. "[My father] was disabled for leaving unto me due maintenance," Dee explained tactfully in a letter to William Cecil, asking for a loan—one of the first of many such supplications.

Dee was himself arrested for treason and heresy by Mary's government in 1555. Although the civil charges were dropped for lack of evidence, he was handed over to the ferocious anti-Protestant bishop Edmund Bonner for religious examination. Bonner, the bishop of London, was conducting his own inquisition, condemning fellow clergymen for failing to conduct Mass, singling out heretics for execution. But it seems that in Dee's case Bonner made an exception and befriended the younger man, ordaining him and making him his chaplain. Dee even lived in Bonner's house for a time. Whether Dee himself participated in Bonner's interrogations and condemnations is an open question. Perhaps he was only teaching the bishop mathematics.

There is no question, however, that Dee was able to maneuver around the issue of religion and survive—and even prosper—during the years of Mary's reign. By January 1556, he felt secure enough to pen a request to the queen for a new project, "Supplication to Q. Mary . . . for the recovery and preservation of ancient writers and monuments." Dee was referring to the state of books and manuscripts in England, which were in danger of being lost forever.

In 1536, on the advice of his closest advisor, Thomas Cromwell, who wanted to reform the bureaucracy and limit the power of the Church, Henry VIII had begun closing down monasteries and appropriating their revenues for the Crown. This policy plumped up the royal coffers so well that within four years *all* monastic property had ceded to the king. Henry used most of the money to finance the army and navy, but some went to reward supporters and some just got lost in the confusion—or stolen. Since the monasteries had been the repository of the country's books, these suffered the same fate as the rest of the goods. As a result, Dee wrote to Mary in his supplication, England was in grave danger of losing its intellectual heritage:

> *Among the exceeding many most lamentable displeasures, that have of late happened unto this realm, through the subverting of religious houses, and the dissolution of other assemblies of godly and learned men, it has been, and for ever, among all learned students, shall be judged, not for the last calamity, the spoile and destruction of so many and so notable libraries, wherein lay the treasures of all Antiquity, and the everlasting seeds of continual excellency in this your Grace's realm.*

Dee then proposed that he go around the countryside, buying up books on the queen's behalf in order to start a "Library Royal."

Mary, who was much more interested in burning those books that spread Protestant doctrine than saving any that might be used

*John Dee in a copy of a seventeenth-century painting* EDGAR FAHS SMITH COLLECTION, UNIVERSITY OF PENNSYLVANIA LIBRARY

to promote scientific research, declined Dee's request. Undaunted, he went ahead and started collecting books on his own. He poked into every corner of England in his pursuit of manuscripts. In his zeal to scoop up ancient texts he often neglected to pay for the books, preferring to borrow them from other collections and then conveniently forgetting to return them. Sometimes he borrowed books without bothering to inform the legitimate owner that he was doing so. He wasn't above taking advantage of his friends' predicaments, either. Some of Sir John Cheke's books ended up in Dee's library while Cheke was off in Antwerp hiding from Mary's agents.

As a result of all of this frenetic book buying and stealing, John Dee accumulated the largest private library in England. By his own count, he had "in all neere 4,000" volumes. By comparison, his alma mater, Cambridge University, had 451 books and manuscripts, and Oxford had even fewer. "No one student, nor any one college, hath a half a dozen of those excellent jewells [books], but the whole stock and store thereof is drawing nigh to utter destruction and extinguishing," Dee lamented. Dee's library continued to be heavily weighted to science, mathematics, and philosophy. He acquired the works of Aristotle, Avicenna, Albertus Magnus, Thomas Aquinas, Robert Grosseteste, Euclid, Ptolemy, and St. Augustine. Also Galen, Hippocrates, Maimonides, Boethius, Alhazen, Plato, and Cicero. He even had the prophesies of Joachim the abbot.

But the cornerstone of his collection was the work of Roger Bacon. Dee saw himself as Bacon's spiritual and intellectual disciple. He had original copies of Bacon's Parisian lectures, his essays on the soul, on animals, and on improving health. He had manuscripts of the *Opus Majus*, the *Opus Tertium*, the *Compendia Studii Philosophii*, a tract on alchemy, another on metals, yet another on codes. He had Bacon's work on *perspectiva* and the calendar, his annotated version of the *Secret of Secrets*, his letter on the power of art and nature and the nullity of magic, another on the utility of arithmetic. Of the 107 works of Roger Bacon now known to have existed during Dee's lifetime—including individual pages and fragments—Dee owned 37. He also owned numerous manuscripts from Bacon's own library.

But Dee did more than collect Bacon. He aggressively promoted Bacon's work and what he saw as Bacon's legacy. Dee annotated the earlier scientist's writings, wrote appreciatively about his methods and ideas, and introduced his work to his friends in the English aristocracy and abroad. At one point, Dee even tried to convince Elizabeth that he was Bacon's direct descendant.

The Roger Bacon whom Dee admired was both scientist and magus. Bacon's emphasis on the primacy of mathematics and its advantage in the field of mechanics bolstered Dee's own theories. There were other similarities: Dee would later undertake Bacon's plan to revise the calendar at the request of Elizabeth's government. Bacon's plea for the study of antique languages in order to be able to read a manuscript void of translation errors made a great deal of sense to Dee, who was teaching himself Hebrew for the purpose of reading the Kaballah.

But the myth of Roger Bacon also appealed to Dee. The Roger Bacon of the brazen head, the Roger Bacon of the magic mirror (Dee had such a mirror himself), the Roger Bacon who wrote in elaborate codes in order to conceal secrets of the utmost importance—that was also the Roger Bacon with whom Dee became fascinated. And so he combed the dark places and haggled with agents and stole—and the work of Roger Bacon was saved.

Then Mary died and Elizabeth ascended the throne.

THE REIGN OF ELIZABETH I OF ENGLAND is one of the most remarkable and best-loved tales in all history. What is not so well known, perhaps, is the nature and complexity of the world surrounding Elizabethan England, as well as the measures taken to ensure her survival. For Elizabeth's rule coincided with one of the most dangerous periods in European history, and for all the queen's attempts to disengage from continental politics, England was nonetheless sucked into the chaos of international affairs. To understand Elizabeth is to understand that she was the only female head of state of the only avowedly Protestant nation in Europe. Sandwiched in between all-powerful Catholic Spain and anguished France like "a bone between two dogs" as one of her councilors

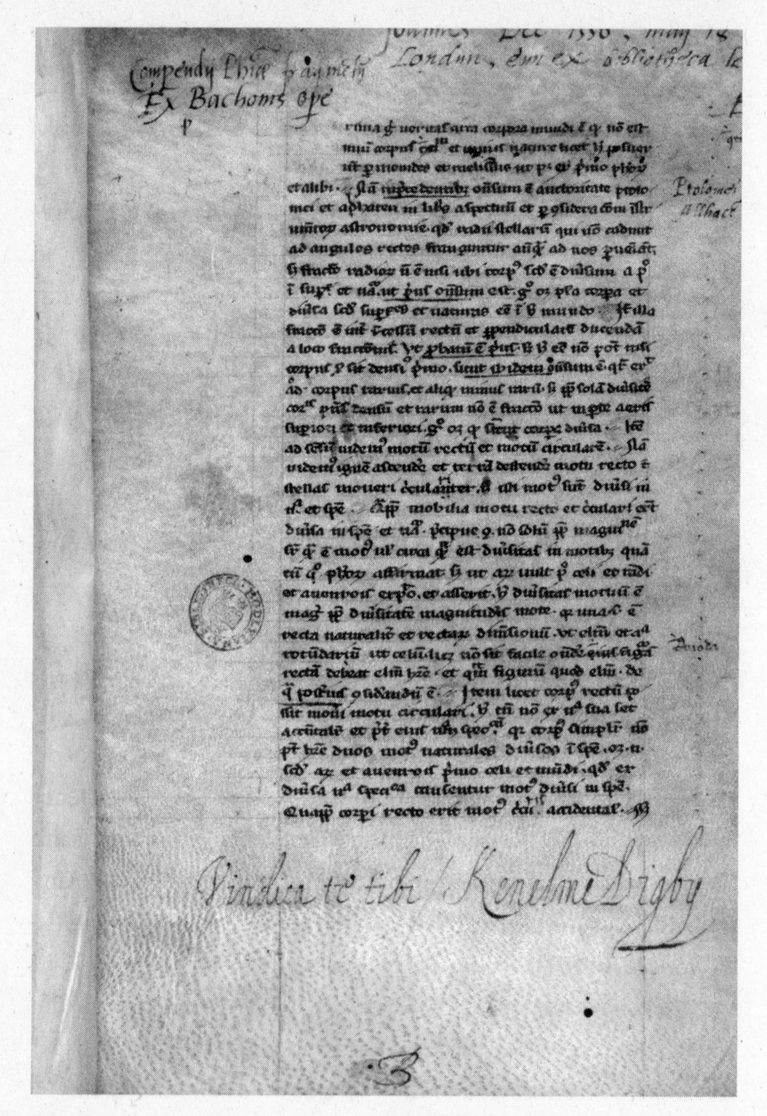

*John Dee's copy of Bacon's* Compendia Studii Philosophii BODLEIAN LIBRARY

put it, Elizabeth had a tenure that marked almost to the day the violent madness of the Catholic initiative to win back territory lost in the previous decades to men like Martin Luther and John Calvin.

The number of plots against Elizabeth during her forty-five-year reign reflects this hazardous state of affairs. For a monarch so beloved, she certainly had a lot of enemies. Even before her ascension, her half sister, Mary, and Mary's ally, Stephen Gardiner, the bishop of Winchester, had thrown Elizabeth into the Tower of London for two harrowing months before confining her to an old hunting lodge in Oxfordshire. (Elizabeth could see the execution scaffolding erected on the Tower Green over the top of the prison wall and wondered daily if she would be the next use of it.) The day after her coronation, her cousin Mary Stuart decided to call herself the queen of England, even though Mary was living in France with her husband, the dauphin. This Mary, known as Queen of Scots (who spoke neither Scottish nor English, having been born and bred a Frenchwoman), would remain a continual threat to Elizabeth. As the granddaughter of Henry VII—Henry VIII's older brother, who died young—she had a legitimate claim to the throne. And this was by no means the end of the list. King Philip of Spain sent his Armada to conquer England and kill Elizabeth; the powerful Guise family in France tried to have her poisoned; and even the pope went so far as to issue a bull excommunicating Elizabeth and promising a spot in heaven to whoever assassinated her.

Uncertain times like these called for innovative measures—and advisors. Elizabeth found hers in Sir Francis Walsingham.

Walsingham was the dark energy of the Elizabethan era. A trusted advisor from 1568, later a member of Elizabeth's privy council, Walsingham almost single-handedly saved the queen from the multitude of villainous intrigues to which she certainly would have otherwise succumbed. In 1568 he wrote to Cecil, "I beseech your honour that I may without offense conclude that in this division that reigneth among us, there is less danger in fearing too much than too little, and that there is nothing more dangerous than security." A devout Protestant and disciple of Machiavelli, Walsingham was committed to his queen and his religion, which he saw as interchangeable. Only a man like Walsingham, preemptive, at once cynical and passionate, could have taken the swirl of Catholic plots surrounding Elizabeth and used them to establish a network of spies and informers that would eventually evolve into the British secret service.

Walsingham's was a world of agents and double agents, of intelligence, counterintelligence, bribes, burns, traps, and moles. He bankrupted his estate by personally financing most of his networks. He began by utilizing the services of one "Franchiotto, the Italian" (a Protestant double agent working for France, who reported that Queen Elizabeth had better be careful about what she ate), and by the time he died in office in 1590 he had half of Europe spying on the other half. He made use of people of both high and low station, Catholic and Protestant alike, although most of his trusted agents were Protestant.

Walsingham understood that in a world of shifting alliances information was power, and so he focused his energy on gathering as much as possible. He had the lord mayor of London prepare reports each week on any suspicious foreigners entering the city and kept tabs on those who stayed. He had ties to Protestant networks in France and Italy from his days abroad and ran agents out of Spain and the Netherlands. He issued fake papers and invented cover stories that allowed his spies to roam freely in enemy territory. He alternately bribed and threatened government officials. He intercepted enemy messages and wasn't above manufacturing incriminating evidence when necessary.

Not all of the threats came from overseas. Although only a small proportion of the English population remained radically committed to Rome and the overthrow of Elizabeth, these people, too, had to be watched. Walsingham was forced by the sheer magnitude of his task to make do with whatever raw material was at hand, and one of those at hand was John Dee.

Dee had emerged miraculously unscarred by his connection to Bishop Bonner and Queen Mary to take up his old position in the royal social circle. Indeed, he was immediately rewarded by Elizabeth with the plum assignment of determining the most auspicious date for her coronation, a sure sign of the new queen's favor and confidence. (By contrast, Bonner was cast into the Marshalsea prison.) John Dee survived because he had been a spy. He had been in correspondence with Pickering and Cecil while they were abroad; he seems even to have been in communication with Elizabeth. The Protestant nobility knew that Mary was ill and that it was only a matter of time before Elizabeth gained the throne. Those who had fled Mary's barbarity were prepared to return to England and begin a new administration as soon as she died. But they needed eyes and ears in England while they were abroad, and Dee was evidently one of those who provided information.

Dee was absolutely committed to Elizabeth. He formed his first attachment to her while she was Mary's prisoner. More than that, Dee had the bourgeois' reverence for title and the trappings of the aristocracy. He loved being a member, however distant, of the queen's society. He kept a diary for the years 1554 to 1601 full of glowing references to the queen and her visits (as well as a minute record of all the money he borrowed over the years):

> *Sept 17th, 1580: the Quene's Majestie cam from Rychemond in her coach, and whan she cam right against the church she turned down toward my howse: and when she was against my garden in the felde she stode there a good while, and than cam ynto the street at the great gate of the felde, where she espyed me at my doore making obeysains to her Majestie; she beckend her hand for me; I cam to her coach side, she very speedily pulled off her glove and gave me her hand to kiss; and to be short, asked me to resort to her court.*

Dee had earlier turned down a position at Oxford teaching mathematics on the grounds that academia would have been too constricting for one who wanted to pursue spiritual mathematics. But what he was really hoping for was a position as court philosopher to Elizabeth.

Elizabeth wasn't a woman to pay for something that she could have for free, however. She got all the advice she wanted but never gave Dee an official position at court. Soon after her coronation, he left on a five-year book-hunting trip to the continent. Dee was to go off and see if he could find anything useful to her administration. If he found something, he would be rewarded.

He did find something. It was a rare manuscript called the *Steganographia* (*On Secret Writing*), written by a fifteenth-century German abbot, Johannes Trithemius. Dee found the book in Antwerp. Its owner, a Hungarian nobleman, lent it to Dee for ten days on condition that the Englishman answer scientific questions on demand. Dee spent ten frantic days copying the manuscript and wrote to Cecil that he had found a book for which others had offered "a Thousand Crownes and yet could not be obteyned . . . a boke for which many a lerned man has long sought and dayly doth seeke; whose use is greater than the fame therof is spread . . . the most precious jewel that I have yet of other men's travails recovered." He then asked for money.

On its surface, the *Steganographia* was a detailed occult work for conjuring up various spirits, a subject in which Dee would have been interested anyway. But actually the book was an instructive manual in the use of highly sophisticated ciphers. Embedded in the incantations were messages based on elaborate codes (alternate letters of alternate words, number substitution, different base languages) nearly impossible to break without a key. Dee, whose knowledge of mathematics and fascination for the dark arts helped make him one of Europe's reigning cryptographers, understood the book's true purpose. His interest in Trithemius's work was threefold. He wanted to use the magic incantations to contact the spiritual world, he wanted to understand the ciphering in the hopes that he could uncover an ancient language based on astrological symbols to unlock the secrets of the universe, and he wanted to prove to Elizabeth's government that ciphers and spiritualism could help protect the queen. When Dee returned to England and moved into a new house called Mortlake, just outside of London, he brought all of his books with him. There was no British Museum at the time. Dee's library became the unofficial athenaeum of the Elizabethan administration. Walsingham, who had a house nearby, was a frequent visitor.

Francis Walsingham was also enormously interested in ciphers, but for more practical purposes. He himself had picked up a book on the subject while in Italy during Mary's reign, and had observed Catholic undercover communications operations first-hand. In fact, ciphers were all the rage in sixteenth-century Europe, because a clandestine operation obviously had a better chance of success if instructions could be passed in an unbreakable code among the various participants. Walsingham needed to develop his own codes and break others. For this he turned to Dee and his library. Dee would certainly have shown him Roger Bacon's work on the subject, as well as briefed him on the *Steganographia*. Dee came up with several ciphers for Walsingham. Unfortunately, most of them were too sophisticated for day-to-day espionage—Dee used alchemical symbols to denote scientific jargon, which in turn was embedded in allegories, so that the encoded messages took a long time to decipher and the communication ran to pages and pages. He did, however, give Walsingham one code that the secretary did use, the "Trellis Cipher," in which the message was read down and then up again rather than from left to right, but Dee sniffed at it as being too simple. It was "a childish cryptogram such as eny man of knowlege shud be able to resolve," he said.

Dee and his library enabled Walsingham not only to develop his own ciphers but to break those used by Elizabeth's enemies. And Elizabeth had no enemy greater than her cousin, Mary, Queen of Scots.

BY 1568, and for the two decades following, Mary, Queen of Scots, had become a serious problem for Elizabeth's administration.

A devout Catholic, Mary wanted to be queen of England. After a scandalous escapade in Scotland, in which her husband murdered her lover and she had another lover murder her husband, she arrived in England, where she would remain for the next twenty years. Her first attempt to seize the throne was to ally herself with the Duke of Norfolk, import troops from abroad, and thereby overthrow Elizabeth.

Two decades later, now under house arrest for her first attempted coup, Mary tried for the throne of England once more. The sophisticated methods used to catch her the second time are testimony to how far Walsingham had come in the ways of espionage and how critical was the use of ciphers to his methods.

In 1583, Elizabeth appointed her cousin Sir Edward Stafford as ambassador to France. The queen never appropriated enough money from government funds to run the place in style, and frequently the ambassador would have to foot the shortfall. Stafford had an even bigger problem—he liked to gamble and ran up big losses. It didn't take long for the Spanish ambassador to France to pick up on Stafford's financial difficulties. "Now was the time for Your Majesty to make use of him [Stafford] if you wished any service done . . . you should see by his acts how willing he was to do so . . . This ambassador is much pressed for money," he wrote to King Philip of Spain. King Philip responded with two thousand crowns. The powerful French Duke of Guise also noticed and paid Stafford's gambling debts as well. By 1585, Stafford had become a spy for both the French and the Spanish, who were conspiring together to unseat Elizabeth and put Mary on the throne.

Walsingham, who had never trusted Stafford (he had a Catholic wife), had spies watching him from the very beginning and soon learned of Stafford's treachery. But since Stafford wasn't especially subtle at espionage, Walsingham left him in place. Walsingham wanted to prove to Elizabeth that Mary was a traitor, so he set a trap.

He had one of his own agents, Thomas Philips, investigate

Mary's household, and sent another agent, a captured Catholic spy named Gilbert Clifford, to the French embassy in London to offer himself as a courier for Mary. Mary and Clifford arranged to have enciphered letters smuggled out in waterproof packages inside the stoppers of the kegs of beer that were routinely delivered to the house to which Mary was confined. The arrangement was that Mary wrote the letters, her secretary transcribed them in ciphertext, the letters went out with the old beer, Clifford took them to the French embassy, the French embassy sent them to Stafford, and Stafford handed them over to the Spanish ambassador.

Unbeknownst to the conspirators, every letter Mary wrote or received went through Walsingham first. By this time Walsingham had an entire cipher department working out of his house in London, with Philips his chief cryptographer. Sometimes Walsingham just read Mary's letters and passed them along as written; sometimes he changed them slightly; and sometimes he kept them and sent his own versions instead.

It was through this unofficial channel that Walsingham learned that the Spanish were planning to build an armada and invade England. So dire was the threat that in 1587 he took the unheard-of step of writing out a covert plan of defense in order to impress upon Elizabeth the urgency of the situation so that she would appropriate funds for his networks. Entitled *Plot for Intelligence out of Spain*, the document outlined a number of surreptitious operations. One of these was to send an agent to Crakow, where Walsingham had a contact who claimed that he could steal the pope's personal correspondence and thereby intercept communications between Spain and the Vatican.

Walsingham's secret agent in Crakow was John Dee.

. . .

DEE HAD NEVER GOTTEN the paid position at court that he wanted and was continually scrounging about for money, but he did whatever Elizabeth asked, no matter how trivial. When she had a toothache and refused to have it pulled, he personally went abroad to converse with the best dentists in Europe for alternate remedies. (Eventually, one of the queen's friends, Bishop Aylmer, had to have one of his own teeth pulled to persuade her to do the same.) Over the years, Walsingham had made use of Dee in various ways. He had dispatched him to France in 1571 to make a horoscope of the Duke d'Anjou, Elizabeth's French suitor (and come back with whatever information he could about French schemes to dethrone the queen). Whenever anybody made a little wax figure of Elizabeth and stuck pins in it, as people sometimes did, Dee was duly called on to mumble incantations and undo the spell. Walsingham sent English explorers and geographers like Richard Hakluyt to Dee for advice and the use of his library, consulted him personally about Admiral Drake's expedition, and encouraged him to find in his old books academic rationales for English territorial expansion. Dee obliged in 1577 by writing two books on the subject, *General and Rare Memorials* and *Brytanici Imperii Limites,* the latter of which included a map of all the lands to which his research supported Elizabethan claims. In so doing, he became the first scholar to formulate the idea of a British empire. "Nov. 28th, [1578], I spake with the Quene hora quinta; I spake with Mr. Secretary Walsingham. I declared to the Quene her title to Greenland, Estetiland and Friseland," Dee recorded in his diary.

In June 1583, Dee had had a visitor, Lord Albert Laski, a Polish prince on an unofficial tour of England. Laski was a Catholic, but as an immense landholder and powerful nobleman he represented an entrée into the court of Rudolph II, king of Bohemia and Hungary. More than that, Laski was ambitious, and ambitious men were useful to Walsingham. Accordingly, the English government went to considerable trouble to make him feel at home. Laski, who was preoccupied with the study of alchemy (land and serfs were all very well, but gold was liquid) and considered himself something of a patron of scientists, expressed an interest in meeting the great English alchemist John Dee, of whose work he had heard so much. Dee noted the meeting in his diary:

> *June 15th [1583], abowt 5 of the clok cam the Polonian Prince Lord Albert Lasky down from Bissham, where he had lodged the night before, being returned from Oxford whither he had gon of purpose to see the universityes, where he was very honorably used and enterteyned. He had in his company Lord Russell, Sir Philip Sidney, and other gentlemen: he was rowed by the Quene's men, he had the barge covered with the Quene's cloth, the Quene's trumpeters, &c. He cam of purpose to do me honor, for which God be praysed!*

Unfortunately, by the time of Laski's visit, John Dee's notion of science and mathematics had descended into the grotesque. A year earlier, he'd made the acquaintance of, and subsequently hired, a man who would exert enormous influence over him for the rest of his life, Edward Kelley.

Kelley had materialized out of the shadows of the Elizabethan underworld one evening to have his horoscope read, and stayed on with Dee, more or less, for the next six turbulent years. On the run from the law on a charge of counterfeiting, he was operating under the assumed name of Talbot when he arrived on the astrologer's doorstep. He had already had his ears lopped off as punishment for his crime and wore a black skullcap over his long, unkempt hair to hide the disfigurement. This gave him a macabre, slightly sinister look that recommended him to Dee, who was by this time engaged in surreptitious acts of magic and was looking around for someone to perform a particularly delicate supernatural task for him. Dee was looking for what the sixteenth century called a "skryer."

A skryer was a spiritual medium, someone who, under the right circumstances, could establish a link between the terrestrial world and that of God and heaven. (There were some skryers who could summon up demons, of course, but Dee, a devoutly religious man, would have nothing to do with them.) In the course of his many foreign travels, Dee had purchased a crystal ball, which he called a magic mirror or "shew-stone." It was one of his most

*Edward Kelley* BEINECKE RARE BOOK AND MANUSCRIPT LIBRARY, YALE UNIVERSITY

prized possessions; he had showed it off to Elizabeth once when she came to visit. But, try as he might, when Dee peered into his rock he saw nothing. Consequently, he was on the lookout for someone to act as intermediary between himself and the angelic spirits. "[I] confessed myself a long time to have been desirous to have help in my philosophical studies through the company and information of the blessed angels of God," Dee wrote. When he mentioned the problem to Kelley, the younger man offered to give it a try. "[Kelley was] willing and desirous to see or show something in spiritual practice."

Kelley was even better at skrying than Dee could have hoped. Kelley saw all kinds of angels in the glass. He saw the angel Uriel, whom God had used to warn Noah of the impending flood. Uriel was a spirit Dee was particularly interested in meeting. He wasn't even an ordinary angel—he was an archangel. After Uriel, Kelley saw Michael, an even more prestigious archangel than Uriel, another on Dee's most wanted list. In fact, through Kelley, Dee got a chance to meet all the really important angels, the ones who could help him in his research.

But Kelley didn't just *see* the angels. He *talked* for them, or they talked through him. They used different voices and different forms. Sometimes they were male and threatening. Sometimes they were flirtatious women. Sometimes they were innocent little girls. Dee would ask them questions and Kelley would answer, roaming around the room, talking in curious, archaic languages, offering advice and revealing secrets, taking on various personalities.

So convinced was Dee that he was indeed in communication with the spirit world that he went to great lengths to ensure that his sessions with Kelley and the angels were productive. He had a whole procedure for it: for three days before, neither man could engage in "Coitus & Gluttony . . . [each must] wash hands, face, cut nails, shave the beard, wash all . . . [prayers should be said] 5 times to the East, as many to the West, so many to the South & to the North." He wrote down every word of these sessions, and they are preserved in his diaries. "It is hard to find in print a more amazing farrago of nonsense," the historian Henry Carrington Bolton noted in his book *The Follies of Science at the Court of Rudolph II.*

By the time of Lord Albert Laski's visit, Kelley was a valued member of Dee's household. But the séances at Mortlake had attracted others' attention as well. Word had got out to the countryside, and Dee was rumored to be a wizard and had become the subject of petty harassments. More important, knowledge of Dee and Kelley's activities had come to Walsingham's attention. A resourceful young scoundrel like Kelley was even more useful to an intelligence network than a credulous old man like Dee. Walsingham blackmailed Kelley into becoming a spy. So it was that when Lord Albert Laski went back to Crakow, he took Dee and Kelley and their wives and children with him.

It is difficult to evaluate their performance as secret agents. Inconspicuous they weren't.

They settled on Laski's estates and proceeded to spend his lordship's money on alchemical experiments at an alarming rate. Laski soon tired of them and sent them off to Prague to the court of Rudolph II with letters of introduction. Rudolph, whose childhood had been spent at the fanatically Catholic court of his uncle, Philip II of Spain, and who had been unbalanced by the experience, was known to have an interest in art and the occult. Alchemy

*Rudolph II of Bohemia*

and the search for the philosopher's stone dominated his life. He kept a fully equipped laboratory manned with alchemists and had once singed his beard while trying to turn lead into gold.

Rudolph had a magnificently idiosyncratic collection of paintings, part masterpiece, part junk, which were hung helter-skelter on the walls of the palace, as well as a library of mystical manuscripts. He kept a menagerie of wild animals, which included a lion cub named Otakar. He maintained huge gardens filled with exotic plants, including flowers brought back from the New World. Rudolph grew the first tulip in Bohemia—he named it Maria, after his mother—and employed many botanists. Although he never married, he fathered many children, and a parade of women went in and out of the castle.

Rudolph, with the treasury of the empire behind him, was known to pay generously for his pleasures. Accordingly, in addition to legitimate scientists and artists—Rudolph supported the astronomer Tycho Brahe and bought paintings by Titian and Holbein—the emperor also attracted a hodgepodge of con men, forgers, and unscrupulous dealers in "curiosities."

Kelley, especially, was in his element. While Dee obtained an audience with the emperor and lectured him on science and mathematics, introducing him to the work of Roger Bacon, Kelley turned lead into gold with the aid of a crucible equipped with a false wax bottom laced with gold filings. Emboldened by success, Kelley staged an even larger demonstration. He invited all of Rudolph's alchemists to witness one of his experiments and had a big box full of instruments and chemicals wheeled into the laboratory. He threw some iron and some red dust, "the philosopher's stone," into the fire and then made everyone leave for an hour while he murmured mystical incantations. When the other scientists had cleared the room, Kelley's brother got out of the false bottom of the box and threw real gold nuggets into the fire.

Kelley's fame thus expanded at the expense of Dee's, and the old man became even more dependent on his protégé. Kelley used his power mercilessly. He would refuse to look into the crystal for Dee, which drove the aging philosopher to despair. Dee tried to get his ten-year-old son, Arthur, to skry for him, but no matter how long the boy sat and stared he never saw anything. When Kelley did condescend to hold a séance, the angels made strange requests. The climax to this absurd drama came when the angels demanded that Dee and Kelley share wives. (Jane, Dee's third wife, was much younger than he was and attractive. By contrast, Mrs. Kelley had long since ceased to interest Mr. Kelley.) Although he demurred at first—there was not much precedent for wife swapping in the Bible—the angels eventually convinced Dee that the command came from God, and he acquiesced.

Through it all, Dee remained in correspondence with Walsingham, and he and Kelley did manage to make contact with the Italian double agent Francesco Pucci. Pucci, a lapsed Catholic, was later accused of having bribed one of the pope's servants to copy letters written by Philip of Spain discussing the sabotage of a certain English forest that supplied timber for the queen's navy. The Spanish were planning to send a small band of men to burn down the forest so that the English would not be able to build new ships. Forewarned by Dee, Walsingham's men were able to intercept this troop before they did any damage.

But Dee also made enemies—Pucci was one of them—and was regarded with suspicion by members of Rudolph's court. It was whispered that he and Kelley were English spies. Kelley, ever the survivor, began to make side deals. The result was that Dee had to flee for his life while Kelley was embraced by the emperor and elevated to knighthood. Before he left, Dee had to surrender most of his valuables to Kelley, ostensibly to be used to bribe Lord Rosenberg, a powerful lord of the realm, for safe conduct. "Feb. 4th [1589], I delivered to Mr. Kelley the powder, the bokes, the glas and the bone, for the Lord Rosenberg; and he thereuppon gave me dischardg in writing of his own hand subscribed and sealed," Dee wrote in his diary.

Dee came back from England to discover that his library had been vandalized in his absence. He lost five hundred volumes, and, although it was claimed that an unruly mob broke in, most of the books ended up in the libraries of other scholars. After his Bohemian adventure, Dee never regained his former influence at court, and his reputation as a scientist and mathematician went into decline. Walsingham died in 1590, still in the queen's service, having secured her safety by subverting the Spanish plans for the

Armada and buying England enough time to prepare for the assault. According to Simon Singh, "After his death, it was discovered that he had been receiving regular reports from twelve locations in France, nine in Germany, four in Spain, four in Italy, and three in the Low Countries, as well as having informants in Constantinople, Algiers and Tripoli."

His employer and protector dead, Dee lived the rest of his life in poverty, even resorting to selling his precious books to live; in the end he was exiled to a nondescript position in Manchester and died in relative obscurity around 1609.

As for Edward Kelley, for three years he was Rudolph's favorite, dubbed the "Golden Knight" for his alchemical triumphs, rewarded with estates and riches. He lived the life of a lord, moved in aristocratic circles, dressed in silks. Then, in 1592, he had an altercation with one of the emperor's retainers, killed his opponent in a duel, and was thrown into a dungeon for the crime. Rudolph thought to use this opportunity to pry out his alchemical secrets and had Kelley tortured, but since the secrets had more to do with subterfuge than science Kelley could not answer his interrogator's questions. In desperation, he tried a daring escape over the castle wall with a rope, fell and sustained internal injuries, and died soon after.

THERE WAS ONE MORE ELEMENT to this story. While in Prague, around the year 1586, Dee's son Arthur remembered his father having a strange book, which he later sold for six hundred gold ducats to the Italian spy Pucci. This book, Arthur later told his friend Thomas Browne, was "a booke . . . containing nothing butt Hieroglyphicks, which booke his father bestowed much time upon: but I could not heare that hee could make it out."

It is just about this time that the emperor Rudolph purchased a manuscript for the hefty sum of six hundred gold ducats. He believed this to be an original work by Roger Bacon, a man in whose magical and scientific achievements he had become extremely interested as a result of his sessions with John Dee. The odd thing about the manuscript was that it was written entirely in cipher.

CHAPTER FIFTEEN

# Brilliant Braggart: Francis Bacon

• • •

IT IS EASY, particularly in the light of the later scandals in Bohemia, to dismiss John Dee's contribution to the advancement of modern science. But the follies of his twilight years and his increasing devotion to spiritualism mask his true legacy. In the fifty years dating from Dee's first commanding lectures on Euclid in Paris, the pace of scientific thought in England, moribund for so many centuries, had quickened perceptibly, so that what had appeared daring and innovative at mid-century was already outdated by 1600. That this was so, that scientific thought was capable of shaking off the shackles of superstition so decisively as to render John Dee's notions ridiculous in his lifetime, was due in no small part to Dee himself—or, rather, to Dee's library at Mortlake.

The library at Mortlake was an engine that helped fuel the scientific revolution in the seventeenth century, of which Britain was an undeniable leader. The presence of so many books, so easily accessible from London and therefore available to so many of England's best minds, jump-started the process of intellectual development in that country in the same way that the sudden influx of translated Greek manuscripts from Toledo had precipitated the scientific renaissance of the thirteenth century. Dee's library succored geographers and mathematicians, philosophers, explorers, and poets. There, at Mortlake, was the embryo for what would eventually grow into the colossus known as the Royal Society.

And since no one could enter that library and not feel the power of Roger Bacon's work, Bacon's thoughts entered the mainstream of English consciousness in a way that would not have been possible otherwise. John Dee had personally promoted the mystique of Roger Bacon, but his library advanced the thirteenth-century friar's ideas. These ideas were absorbed by many scholars, but one in particular stands out as Bacon's true intellectual descendant, the man who finally brought Roger Bacon's notion of experimental science into the sunlight, so much so that because they hold the same surname, the two are often confused: Francis Bacon.

*John Dee performing an alchemical feat for Rudolph II in a nineteenth-century painting* EDGAR FAHS SMITH COLLECTION, UNIVERSITY OF PENNSYLVANIA LIBRARY

. . .

FEW IN THE HISTORY OF PHILOSOPHY OR SCIENCE have a larger presence than Francis Bacon, but the question is, as what? Genius, humbug, solitary scholar, stentorious self-promoter, loyal friend, ambitious betrayer, judge, thief, social climber, member of Parliament, mystic, experimentalist, original thinker, shameless plagiarist, and sometime reputed author of Shakespeare's plays—Francis Bacon was all of these and more. His proponents have pronounced him the father of modern science; detractors asserted that "few attempts at giving a new direction to the pursuit of truth have been more overrated." Will Durant described him as "the greatest and proudest intellect of [his] age"; the social anthropologist Loren Eiseley called him "the man who saw through time"; and the scientific historian Lynn Thorndike said that "he was a crooked chancellor in a moral sense and a crooked naturalist in an intellectual and scientific sense."

Francis Bacon was born in 1561, three years after Elizabeth ascended the throne. Like Roger Bacon, he was a younger son; his father, Nicholas, was the keeper of the royal seal, a position of great prestige. The most important member of the family was his uncle by marriage, the even more prestigious (and extremely wealthy) William Cecil, Lord Burghley, Elizabeth's royal treasurer and first secretary of state.

When he was twelve, Francis was shipped off to Cambridge for much the same reasons as had been Roger Bacon to Oxford 350 years before. When his father died six years later, Francis, by then in the queen's service, saw that his road for both position and money went through Cecil. But Cecil did not think much of his egocentric and crassly ambitious nephew. In one of his many entreaties to his uncle to provide a path to position and wealth that went unanswered, Bacon is famous for having said, "The objection of my years will wear away the length of my suit."

The life of a gentleman seemingly beyond reach, Bacon turned to practical pursuits. He earned a law degree in 1582, showing such aptitude for the profession that he was soon lecturing, and then at age twenty-three got himself elected to Parliament, where he would remain a member for almost forty years. From here, Bacon spent much of the next decade currying favor with the powerful while serving in a series of minor roles in English government. In his free time, he began serious study of history, philosophy, and science. "I have taken all knowledge to be my province," he wrote to Cecil, with the utter lack of humility that had already become his trademark, doing little to dispel his uncle's opinion of him.

Like every other scholar of the period, Bacon spent little time at the largely denuded university libraries. Rather, he sought out the best-stocked of the private libraries, and the largest and most extensive personal collection in England was at Mortlake.

Nicholas Bacon had been a friend of Dee's, and there is an entry in Dee's diary for August 11, 1582, indicating that twenty-one-year-old Francis had visited him and made use of his books. (Also present at that meeting was Walsingham's cryptographer, Thomas Philips.) There is no specific documentation of further visits, but it has long been suspected that Dee provided the young, prodigious Francis Bacon a detailed introduction to the work of his thirteenth-century namesake, even mentored him in experimentalism. Bacon's subsequent output in the sciences would more than bear that out. It was shortly after the Mortlake visit—or visits—that Bacon began work on a vast scientific project on which he would labor for the rest of his life. It was also about this time that he developed a lifelong interest in ciphers.*

It would be more than two decades, however, before Bacon would write publicly on science. During Elizabeth's reign, he wrote mostly on politics and ethics, showing such foresight that a seventeenth-century observer noted that, had his ideas been adopted, "all the troubles of the next forty years might have been avoided." Foresight did not make Bacon popular at court, however. Convinced that unity would serve England more than discord, he issued a paper in favor of religious tolerance that piqued both the queen and Cecil. Both made it clear to Bacon that if he wished to advance himself—a desire of which he made little secret—he was going to have to keep his mouth shut and play along. Bacon, however, was unwilling to suppress opinions that he was convinced were the most enlightened in England (and probably anywhere else). In 1593, to demonstrate his unwillingness to kowtow to the aging Cecil or the intractable Elizabeth, he publicly declared himself opposed to the queen's plan to raise taxes.

Elizabeth was furious, and Bacon was shunned at court. The

* Bacon's cryptology is a major factor in the "Shakespeare controversy." According to proponents, what appear to be plays by the illiterate actor William Shakespeare were actually the work of Francis Bacon. Bacon is known to have created what he called the "biliteral cipher," whose key is based on the use of two different typefaces, what are today called "fonts." By varying combinations of the two typefaces, words in the text of any printed work (one of Shakespeare's plays, for instance) could be used to embed secret messages and cipher signatures. The cipher works by assigning every letter of the alphabet (the plaintext) a unique combination of typefaces "a" and "b"—say, italics and Roman—in groups of five letters. To decrypt the cipher, the recipient would simply note the order in which the typefaces appear. So, a "d" in plaintext would be represented by a group of letters with typefaces in an arrangement of "*aaa*bb," or "*italics, italics, italics,* Roman, Roman," as in "*hel*lo." It doesn't matter what the letters are in the text, only in which typeface they appear. It doesn't take much to see that this can lead to some highly subjective decryptions. While there are some diehards even today who cling to the belief that Francis Bacon actually wrote Shakespeare's plays, the notion was effectively demolished in 1957 by two brilliant cryptanalysts who will later play a significant role in this story.

direct road to power now cut off, Bacon was forced to cast about for a new sponsor. He found one in Robert Devereux, Second Earl of Essex, a man of his own generation and disposition, and there is nothing that epitomizes the extremes of Bacon's public life and reputation more than the course of their friendship.

Essex, five years Bacon's junior, was handsome, brave, intelligent, headstrong, passionate, and a favorite of the queen's, even rumored to have been her lover. (Errol Flynn played him in the movies, too, in *Elizabeth and Essex*.) He became both Bacon's friend and patron, while in turn Bacon provided counsel as to how Essex might increase his power in the government (and by association, Bacon's own), something the ambitious earl was chafing to do. In the early 1590s, the grateful Essex pushed for Bacon to be named attorney general, a plan foiled by Cecil himself, who instead prevailed upon Elizabeth to appoint the more agreeable Edward Coke. Essex consoled Bacon by granting him an estate at Twickenham and a stipend of £1,800 to keep it up.

In 1598, Cecil finally died, and, once again largely through the efforts of Essex, Bacon was appointed Queen's Counsel, part of a group of lawyers who, in addition to their judicial tasks, advised the Privy Council. Soon after, Essex committed a series of political blunders (all against Bacon's advice) and, almost impossibly, succeeded in alienating Elizabeth's affection. Finally, in 1601, he joined in a plot (with Mary's son, James VI of Scotland, among others) to raise an army and take over the government. When he was caught, it was Bacon who was charged with drawing up the indictment. Bacon, torn between political suicide and betraying his closest friend, chose the latter. Not only did he present the indictment but, when Coke appeared to be botching the job at the trial, pled the case himself. Essex, proud and gallant until the end, freely confessed his guilt and a few days later was executed on Tower Hill. His severed head remained on public display for a year.

If Bacon's ambition and ego were colossal, so was his capacity for achievement, and it was only after Elizabeth's death and the ascension to the throne of Essex's collaborator James VI (now James I of England) that all of these would find their outlets.

As soon as the new king was in place, Bacon wrote a long, fawning letter proposing a brilliant candidate for high public office—himself. James, not altogether enamored of the man who had prosecuted Essex, ignored the suggestion but did throw the supplicant a bone by including Bacon in a list of three hundred new knights. Undeterred, the now Sir Francis continued to pester the king with missives, suggestions for improving government, stressing his favored themes of ethics, unity of purpose, and conciliation. For all the erudition, however, Bacon continued to be disregarded.

Then, in 1605, Francis Bacon published the first of his great scientific works. Much as Roger Bacon had proposed that Clement reform education and rebuild the trivium and quadrivium from the ground up, this work, which Francis Bacon called *On the Dignity and Advancement of Learning,* was a proposal to completely redefine the role and manner of education in English society.

As Roger Bacon began with a tribute to the pope, Francis Bacon began with one to the king. Taking no chances, he made this tribute even more obsequious than those before it. "I have been often struck with admiration, apart from your other gifts of virtue and fortune, at the surprising development of that part of your nature which philosophers call intellectual," he wrote to a king considered not particularly bright. "The deep and broad capacity of your mind, the grasp of your memory, the quickness of your apprehension, the penetration of your judgment, your lucid method of arrangement, and easy facility of speech . . ."

Bacon then noted, "Since your majesty surpasses other monarchs by this property [the learning of a philosopher] . . . it is but just that this dignified pre-eminence should not only be celebrated in the mouths of the present age . . . but also that it should be *engraved in some solid work which might serve to denote the power of so great a king and the height of his learning*" (italics added).

All that groveling notwithstanding, what followed was truly groundbreaking—*The Advancement of Learning* was as close as anyone in Europe had come to calling for universal education. Bacon insisted that learning be pursued both for its own sake and for its practical benefits to society. Although he never actually wrote "knowledge is power," a saying for which he is most famous, he was to later state that "knowledge and human power are synonymous." This sense of practical utility was precisely what Roger Bacon had stressed so strongly to Clement in his plea for the reform of scholasticism. He had promised Clement military, civil, and even ecclesiastical benefits from a more empirical view of the world—a stronger and more vibrant Christendom—and so the more secular Francis Bacon assured James that the spread of learning would result in a stronger and more vibrant England.

There would be objections to be overcome, however—some sincere, some self-serving, but all misguided—and once more Francis Bacon looked to the same culprits as had Roger. Instead of the tyrannical Dominicans whom Roger Bacon blamed for the perpetuation of ignorance, Francis Bacon denounced "arrogant politicians" and "zealous divines," but the litany of excuses for an uninformed populace was the same—that educated men would not make good or willing soldiers (or churchgoers), that they would be less accepting of the rules of king or Parliament (or pope), that it would make people indolent and slothful (or questioning).

In the second part of *Advancement of Learning,* Bacon provided a detailed breakdown of all possible areas of inquiry, noting which had been explored and which needed further study. As with the *Opus Majus, Advancement of Learning* stressed the study of empirical science in the pursuit of general knowledge. Physics, experimentation, invention, medicine, natural history, anthropology, botany—not only are all of these mentioned prominently, but many are singled out by Bacon as those areas where further investigation is most warranted.

His one major break was in the discipline that the earlier Bacon had called "the gate and key of knowledge." In *Advancement of Learning,* mathematics was cataloged not under science but in "metaphysics," a category that Francis Bacon dismissed as mere philosophy and not part of scientific inquiry. "Quantity," which he considered the sole subject of mathematics (a much narrower definition than that found in the *Opus Majus*), was "abstract and separable from matter," and anything that was abstract detracted from direct study of the physical world. "Mathematicians," Bacon wrote, "would have their science [and logic] preside over physics," instead of assuming the proper role as "handmaids to physics." This lifelong prejudice against mathematics would prove to be one of the most controversial aspects of Bacon's scientific theories and, as experiments became more and more sophisticated, one of its greatest limitations.

Although none of Francis Bacon's proposals was put into practice, after the publication of *Advancement of Learning* the king was won over—whether by argument or flattery—and from there

***Francis Bacon pictured in a frontispiece of a 1640 edition of* Advancement of Learning** EDGAR FAHS SMITH COLLECTION, UNIVERSITY OF PENNSYLVANIA LIBRARY

Bacon's rise was meteoric. He was named solicitor general in 1607, in 1613 finally became attorney general, three years later was appointed to the Privy Council, the following year granted his father's old job as lord keeper, then finally, the year after that, 1617, was appointed lord chancellor. His private fortunes improved as well. In 1606, despite his personal dislike of women and children and fondness for young men, he married a wealthy widow, Alice Barnham, guaranteeing himself an income of £220 per year. (He never had children, whom he had termed "hostages to fortune," a fitting description from a man left penniless by his father.) In 1618, he was made Baron Verulam and then in 1621, Viscount St. Albans.

Bacon used his influence to try to advance his political agenda, which, in today's terms, would be called "engagement." He believed that deft use of diplomacy and limited conciliation could overcome friction between the religions, the two houses of Parliament, and the disinclination of both Wales and Scotland to be part of a greater England. Although James ignored Bacon's advice as excessively radical, it had little effect on the esteem with which he was now held at court. When the king journeyed to Scotland, it was to Bacon that he left the responsibility of ruling England.

In 1620, at the height of his influence, Bacon announced plans for a great six-part encyclopedia, which he dubbed *Instauratio Magna*, or "The Great Renewal." One part was to be *Advancement of Learning*. Another was his proposal for a revolutionary approach to science, to be entitled *Novum Organum*, a direct swipe at Aristotle's logical works, the *Organon*, which Bacon sought to replace and "begin the entire labor of the mind again." In the preface to the *Instauratio* he wrote, "Francis of Verulam reasoned thus with himself, and judged it to be in the interest of the present and future generations that they should be made acquainted with his thoughts."

But how much of what followed were his thoughts? Bacon is notorious for his unattributed "borrowings," and in this work, even a casual observer can readily see shades of almost every scientific theorist who preceded him, but there is no one from whom

Bacon pirated more and in such blatantly specific detail than his namesake. Roger Bacon is everywhere in Francis Bacon's scientific work—the idea for the encyclopedia itself is more than vaguely reminiscent the *Scriptum Principali*—but there is no more egregious example than in the *Novum Organum.*

Here, Roger Bacon's four causes of error, "obstacles to grasping truth," have morphed into Francis Bacon's "four species of idols [that] beset the human mind." Roger Bacon's four, it will be remembered, were "submission to faulty or unworthy authority; influence of custom; popular prejudice; and, concealment of our own ignorance accompanied by an ostentatious display of our knowledge." The latter Bacon, never willing to miss an opportunity for florid prose, calls these "idols of the tribe," "idols of the den," "idols of the marketplace," and "idols of the theater."

"Idols of the tribe" are defined as distortions of perception inherent in human nature, that "all perceptions both of the senses and the mind bear reference to the man and not the universe," a concept that might easily be seen as popular prejudice. "Idols of the den" are those distortions of perception particular to each individual, acquired from his "education and intercourse with others, or from his reading and the authority acquired by those whom he reveres and admires." It would be hard to get closer to submission to unworthy authority than that. "Idols of the market" are "formed by the reciprocal intercourse and society of man with man . . . from the commerce and association of men with each other." In other words, popular prejudice.

His fourth idol, that of the theater, came about through the perpetuation of "peculiar systems of philosophy and perverted rules of demonstration . . . creating fictitious or theatrical worlds." Where Roger Bacon spoke of this error from the point of view of

***Frontispiece of the* Instauratio Magna *showing a ship traveling between the Pillars of Hercules, believed to mark the end of the known world, an allusion to the new knowledge Bacon was bestowing upon the human race*** **BEINECKE RARE BOOK AND MANUSCRIPT LIBRARY, YALE UNIVERSITY**

those who perpetrated it—his Dominican adversaries in particular—Francis Bacon discusses its effect on the listener.

As to the science itself, in order to "begin the entire labor of the mind again," Bacon began by throwing out both Aristotle and his deductive method. There was no greater source of corruption of both science and human thought, he insisted, than the syllogism (all Aristotelian conclusions are correct; that the universe is eternal is an Aristotelian conclusion; therefore the universe is eternal). Hypothesis—assumption—the essence of propositions in a syllogism, cannot be a means to truth, since the very process of making an assumption limits and may misdirect any conclusions that might be drawn from it. He denounced scholastics as prime culprits in deductive miscreancy—"the corruption of philosophy by the mixing of it up with superstition and theology," as he put it.

The answer, Bacon proposed, was to abandon deduction in favor of induction—the slow road to knowledge of general principles from pieces of verifiable evidence. Francis Bacon is so renowned for his furious criticism of Aristotle and championing of induction that he is often referred to as the inventor of the inductive method. The only problem, of course, is that the Philosopher himself in his work on the theory of knowledge, *Posterior Analytics*, specifically mentioned induction as one of the two methods of acquiring knowledge. Bacon acknowledged that Aristotle occasionally presented a perverted form of induction but claimed that "true induction"—Baconian induction—was similar in name only. A scientific record "compiled on its own account [Aristotle's *On Animals*, for example]," he wrote, "and one collected for the mind's information as a foundation for philosophy [his own method] are two different things."

Bacon's new and original recipe for induction, which relied heavily on Roger Bacon's *Scientia Experimentalis,* was perhaps the most rigorous and elegant blueprint for scientific inquiry ever devised. His aim was to create what he called a "natural history" for phenomena in the observable and measurable world through a series of observations. Natural histories could be created not just for living creatures or rocks but also for events such as heat, light, or astronomical motion. The observations he proposed were not to be random, but controlled, measured, and then recorded—in other words, experiments. "The secrets of nature betray themselves more readily when tormented by art [experiment] than when left to their own course," he wrote. Progress toward general truth could occur only when "numerous experiments shall be received and collected." In an insight that has become the basis of modern theoretical physics, Bacon noted that the principal value of experiment was its ability to disprove a spurious theory, not to confirm a truth previously held.

But not all experiments advanced truth. Bacon separated those experiments that merely had some utility to society—or could result in a profit—from those that moved science closer to a general truth. Also, the scientist must not "jump and fly from particulars to remote and most general axioms," but rather move slowly, from one small intermediate step to another, forming tentative axioms—working hypotheses—only when the weight of experiment allowed. Only then would he begin to approach an understanding of the general, the "rule."

Again like Roger Bacon, Francis Bacon insisted that science be studied in order to be used. The goal of science was invention. He noted that three inventions in particular, the printing press, gunpowder, and the magnet, had "changed the whole face and state of things around the world." Science like education would create a stronger, happier, more patriotic, and even a more pious culture.

In addition to shunning mathematics, Baconian science had some other obvious drawbacks. Although he might have been correct in distrusting hypothesis and deduction, he was unable to dispense with either. In both the delineation of areas of study in

*Advancement of Learning* and the experimental method in *Novum Organum*, Bacon employed the same complex system of classification and subclassification for which he denounced Aristotle and relied just as much on assumption. For all the verbiage, there was little difference between them. What, after all, is the classification of mathematics as a branch of metaphysics except an extension of his assumption that abstract thought is inherently unscientific? Also, Bacon needed hypothesis to resolve conflicting experimental results.

All of this bothered him not at all. "The die is cast," he wrote, "the book is written, to be read either now or by posterity—I care not which; it may wait a century for a reader, as God has waited 6000 years for an observer."

Bacon did not have to wait nearly that long for yet another reversal of fortune. In 1621, disaster struck once more. In his legal affairs, Bacon had been a productive, highly competent judge, noted for his lucid, incisive verdicts and his industry in clearing huge court backlogs. But the preeminent position at court that he had so assiduously pursued became his undoing. During a dispute between Parliament and the king, Bacon became the perfect surrogate to attack. Led by his old nemesis, Edward Coke, Bacon was accused of taking bribes on the bench. Accepting money from participants in a trial was about as uncommon as gambling in Rick's Café, but James's enemies demanded action all the same. Bacon, while freely admitting that he had taken money, pointed out that in the two cases at issue he had ruled against the gift givers.

Nonetheless, James was forced to remove Bacon from all his official posts, fine him £40,000 (which was never paid), and sentence Bacon to the Tower (in which he spent only three days). Professionally, Bacon was ruined, and his reputation has never recovered. He retired from the public sphere and spent the last five years of his remarkable life engrossed in his studies.

Francis Bacon died totally in character. During a carriage ride in the country on a frosty day in early April 1626, he looked out to the roadside and it occurred to him to conduct an experiment. To try to determine if snow could be as effective a preservative as salt, he stopped to buy a fowl, had it killed, then stuffed it with the snow to see if and for how long it could prevent the bird's flesh from rotting. The bird was preserved, but not Bacon. He caught a chill and was brought to the home of a friend, Lord Arundel (who was imprisoned in the Tower at the time for a marital indiscretion). Placed by the servants in a damp bed, he developed a fever, and a few days later on April 9, 1626, Francis Bacon died, probably of pneumonia. In his final letter, he wrote to thank Arundel for the hospitality of his house and to recount the circumstances of his illness. "As for the experiment itself," he added, "it succeeded excellently well."

Even in death, however, Bacon refused to be silent, and one year later he produced one of his most influential works. (It had probably been written in 1624.) Entitled *The New Atlantis*, it was a utopian fable that combined a bit of Shakespeare, a bit of Thomas More, a dash of Plato (whom Bacon also supposedly hated), and, once more, a healthy dollop of Roger Bacon.

The story, such as it was, involved a group of travelers who were becalmed and threatened with starvation somewhere in the South Pacific off the coast of Peru (a fairly exotic locale in 1627). They came upon a mysterious, heretofore unknown island that was home to a reclusive civilization living in peace, tranquility, and with religious and ethnic tolerance. (Many of the plot elements of *The New Atlantis* are remarkably reminiscent of James Hilton's classic 1933 novel, *Lost Horizon*.)

The central institution of this civilization—and the raison d'être of the story—was the "House of Salomon," essentially a vast experimental laboratory run by a group of monklike wise men who were reclusive even from the ordinary residents of the island. The House of Salomon was a paean to applied science, producing wonders of every sort. There were caves and tunnels deep under the earth to study refrigeration, preservation, and the making of metallic alloys, and towers half a mile high to investigate wind, lightning, and other natural phenomena (although not astronomical bodies). Experiments had produced medicinal plants, superior flora (by grafting) and fauna (by cross-breeding), and water infused with minerals to prolong life. Food was plentiful and nutritious and creature comforts abundant and available to all.

There were also a number of mechanical inventions that were similar to those enunciated by Roger Bacon in his letter of 1248. The House of Salomon had produced both telescopes and microscopes (rudimentary prototypes of which were by then available in Europe), prisms, crystals, advanced weapons, superior gunpowder, motorized cars, flying machines, and submarines.

Although there had been a number of works like *The New Atlantis* published in the years before, the idea of an institution devoted to the advancement of science had never been presented quite this way, and it fascinated English intellectuals. The House of Salomon became the model of a number of clubs that sprang up in London, Oxford, and Cambridge.

In 1645, one of these clubs began weekly meetings in London, its purpose to explore science in the Baconian tradition. Each of the ten members was charged a shilling to help defray the costs of materials needed for their experiments. One of the members later wrote,

> *We barred all Discourses of Divinity, of State-Affairs, and of News . . . confining ourselves to Philosophical Inquiries . . . such as . . . Physick [medicine], Anatomy, Geometry, Astronomy, Navigation, Staticks, Mechanicks, and Natural Experiments. We there discoursed the Circulation of the Blood, the Valves in the Veins, the Copernican Hypothesis, the Nature of Comets, the new Stars, the Attendants on Jupiter, the Oval shape of Saturn, the inequalities and Selenography of the Moon . . . the Grinding of Glasses [lenses], the Weight of Air, the Impossibility of Vacuities and Nature's abhorrence thereof.*

Of the ten original members, it was the one who was forced to miss the first meeting who was to be the most important: Robert Boyle, considered the father of modern chemistry. Eventually, Boyle and a number of his fellows would receive a charter from the newly restored Charles II and found the Royal Society of London for Improving Natural Knowledge. Its membership would soon include some of the greatest names in the history of science—Boyle, Robert Hooke, Christopher Wren, Edmund Halley, and, of course, Isaac Newton. From the first, the Royal Society was a group dedicated to the type of scientific inquiry about which Francis Bacon had been so fervent, although hypothesis and deduction were employed every bit as much as (if not more than) Bacon's beloved induction. Still, as Joseph Glanvill, one of the Royal Society's earliest members, noted, "Salomon's House in the *NEW ATLANTIS* was a Prophetick Sceam of the *ROYAL SOCIETY.*"

***Frontispiece from Sprat's* History of the Royal Society, *London 1667, with Francis Bacon sitting to the left of a bust of Charles II*** BEINECKE RARE BOOK AND MANUSCRIPT LIBRARY, YALE UNIVERSITY

Thomas Sprat, who wrote in 1667 of the founding of the society, said, "There should have been no preface to my account of the Royal Society but some of [Bacon's] writings." Boyle himself added that Bacon was "the first among the moderns who handled the doctrine of heat like an experimental philosopher."

LYNN THORNDIKE DISMISSED FRANCIS BACON'S WORK by comparing it unfavorably with that of someone else about whom he had nothing good to say. "It was a relatively easy thing to criticize the past and present state of learning, and to advocate a new program using 'experimental science.' Roger Bacon had done it three and a half centuries before."

Francis Bacon was certainly an easy man to dislike. His ego may well have outstripped his talents, his science was imperfect, and some of his blunders today look laughable. (He rejected Copernican astronomy with the same off-hand disdain as every other theory that he thought beneath his methods.) More than that, he was often forced to rely on the very techniques he denounced—he regularly formed hypotheses and employed deduction whenever it suited him—and proposed experiments that had already been performed. Additionally, he never attempted to reconcile inconsistencies in his theory.

But all of this misses the point, much as Thorndike missed the point with Roger Bacon. Francis Bacon was insatiably curious and willing—in fact, eager—to challenge any piece of evidence that he deemed spurious or obtained by incorrect method. That he drew freely on the work of others without acknowledgment is undeniable, but, in the end, how much did that really matter? Originality does not always determine effectiveness. Would Bacon's contribution to the advancement of human knowledge have been somehow greater for the presence of footnotes?

It is not always the most admirable of characters who change history. For all his smugness, superiority, and passing off others' ideas as his own, Francis Bacon succeeded in achieving that which Roger Bacon tried so desperately but failed to attain. Be it as a result of timing, luck, or simply the ability to couch ideas as his contemporaries needed to hear them, Francis Bacon—deserving or not—became the spirit of a new scientific revolution. He was the fulcrum upon which the balance of science and theology would pivot and that allowed others about whom there is no ambiguity—Boyle, Hooke, Newton, Halley—to tip that balance permanently. It was Bacon's methodology as much as any other single factor that was responsible for finally discrediting scholasticism as a basis of science, so that, after centuries of elephantine progress, Thomas Aquinas could now be relegated to theology only, never again to be taken seriously in the pursuits of science.

CHAPTER SIXTEEN

# The Trail of the Cipher Manuscript

. . .

THE CIPHER MANUSCRIPT ACQUIRED BY EMPEROR RUDOLPH for six hundred ducats was a not-particularly-grand volume of something over two hundred pages. The cover was vellum but flimsy and blank, making the book look more like someone's personal journal than a finished publication. The writing, which looked like Latin but wasn't, varied between a more open script and a smaller, more delicate hand; similarly, the illustrations were sometimes broadly drawn, sometimes finely detailed, as though two different scribes had been employed to copy the work.

Although the writing was mysterious, it was the illustrations that were the manuscript's chief attraction. The entire first sixty pages or so were obviously dedicated to botany, with a specimen drawn on every page, but no actual medieval or renaissance garden had ever held plants like these. They were voluptuous, fantastic; the root systems defied classification; in one illustration the last tendrils turned into faces. The pictures were obviously integral to the text and almost contemporary in style. They were all muted watercolors with none of the rich cobalt blues and gold leafing of a typical illustrated manuscript.

After the plants came an astrological section with large, fold-out pages, clearly the heart of the work. Here were Mars and Taurus and other recognizable astrological figures ringed by stars and spheres as were commonplace in the sixteenth century, but the similarity ended there: inside the rings were rough drawings of rotund naked women bathing in fountains and barrels, wearing expressions of coy surprise, as if the water was too cold. There were also intricately drawn maps of no immediately recognizable area.

Following the astrological section, the manuscript reverted to the strange exotic plants, with the significant addition that this time the little naked women were *inside* the plants, sometimes a string of them in the leaves, which seemed to equate them with the plants, leaving the overall impression of fertility. Some of the plants and women were labeled with the same cipher-word. There was one more difference: if one looked very closely, one could see that some of the women were actually men.

The last section had no drawings at all. It seemed to be a glossary or index, perhaps a dictionary of terms, like a pharmaceutical guide.

The manuscript could not have more perfectly suited the emperor's tastes or excited his interest. Not only had John Dee introduced Rudolph to Roger Bacon's work, he had communicated his own obsession with the thirteenth-century scientist to the monarch. He had spoken to Rudolph of Roger Bacon's encyclopedic knowledge, of his commitment to learning all that could be known of science, of the mysteries he had discovered and hidden in cipher. He discoursed on optics and mathematics and talked knowledgeably of submarines, flying machines, burning lenses, talking heads, and gunpowder. He told Rudolph of Bacon's interest in medicine and health and of Dee's own search for the philosopher's stone, which would unlock the secrets of the universe and eternal life. Eternal life was a topic very near to Rudolph's heart—he was keen to experience it himself. So, naturally, when a manuscript turned up in a strange language that no one could read except for a three-line cipher at the end attributing the work to Roger Bacon, Rudolph bought it. He paid handsomely for it, too—the equivalent of $100,000 today.

Was this, perhaps, the missing section of the great *Scriptum Principali* that Bacon had begun but supposedly never finished? Was this what he had worked on secretly all those years while imprisoned by the Franciscans, when a man known for the volume of his output had apparently and unaccountably shut down? That might account for the complexity of the cipher—in case of discovery, his jailers would not have been able to read the work, and that would, at least, limit additional charges of heresy. On the other hand, although Bacon had written on health and longevity, he was neither a botanist nor a physician. Erotic drawings of foliage and little naked women had never appeared in one of his manuscripts before. Then again, who knew what fifteen years of imprisonment on bread and water would do to a person?

It is not known what Rudolph initially did with his new purchase, but sometime between 1608 and 1616 it came into the possession of his chief botanist, Jacobus Horcicky de Tepenec, also known as Sinapius.

There are two versions to the story of Horcicky's life, one decidedly more romantic than the other. The more prosaic account has Jacobus born poor, of peasant ancestry, in southern Bohemia in 1575. His family gave him to the Jesuits as a menial in exchange for board, but his obvious intellectual abilities brought him quickly to the attention of the rector, who arranged to have him attend the local school in 1590.

Jacobus did very well at his studies and was particularly drawn to the school apothecary, an experienced druggist who concocted and sold various potions for the aid of the sick. Since much of sixteenth- and seventeenth-century medicine was based on herbal remedies, Horcicky's interest inevitably drew him to the study of plants. He graduated at the top of his class, and the Jesuits were so impressed with his performance that they sent him on to the university at Prague for training in Aristotelian philosophy.

In Prague, Jacobus followed the scholarly tradition of Latinizing his name and became Sinapius. He apparently tried to study Aristotle but didn't have the knack for it and instead persuaded his teachers to let him work in the university gardens and set up a laboratory for extracting medicines and other herbal remedies. He became a skilled botanist and chemist. He struck pay dirt when he extracted the essence of flowers and invented toilet water, which he marketed under the label "Aqua Sinapii." At a time when few bathed regularly, Aqua Sinapii was extremely popular and made its inventor a large fortune. He became so rich that Rudolph, hearing of his success, borrowed money from him, signing over the prosperous district of Melnick in return. In 1607, Rudolph brought him to court as chief apothecary and head of the imperial gardens. The next year, one of Horcicky's potions was credited with saving Rudolph's life. In gratitude, Rudolph made him noble, giving him the title "de Tepenec."

In the second, more romantic version of Horcicky's life, Jacobus was born in Prague, the son of a famous apothecary, Christian Horcicky. Christian owned the "City Pharmacy," the most

*Jacobus Horcicky de Tepenec* COURTESY RAFAL PRINKE

successful venture of its kind in all of Bohemia. In this account, Jacobus came by his profession naturally, growing up amid the bottles of elixirs and turpentine, the wooden crates of myrrh, saffron, ginger, and wormwood, and the jars of crabs' eyes, asses' hooves, and dried toads that were the sixteenth- and -seventeenth-century pharmaceutical stock in trade. The City Pharmacy sold everything from cosmetics to sealing wax and was so well known that as a young man Jacobus was assigned to oversee the planting of the imperial gardens, where he was able to examine exotic foliage from all over the world. Father and son together invented the Aqua Sinapii that made the family fortune, Sinapii again being the Latinized form of Horcicky, which incidentally meant "mustard" in Bohemia.

Jacobus had a pretty sister, Sofie, and a good friend, Lieutenant Maximillian Swoboda, an officer in the imperial cavalry. Sofie and Swoboda were in love and engaged to be married. Unfortunately, Sofie was also desired by one Carlo Malombra, an Italian refugee working as an apprentice in the Horcicky laboratory. Malombra, diminutive and squat, with one leg shorter than the other, had been a medical student at the University of Padua but had been forced to flee when accused of poisoning an enemy with a particularly toxic potion known as "Aqua Toffnina." Malombra resented his lowly position in Bohemia with the Horcicky family and dreamed of marrying the daughter of the house.

To get rid of Swoboda, Malombra poisoned the oats in the royal stable and framed his rival for the crime. Two of Rudolph's favorite horses were found dead one morning, and the only clue was a silver spur marked with an *S*, which had conveniently been left at the scene. Lieutenant Swoboda was accordingly arrested for the heinous deed and thrown into jail pending trial. The Horcicky family was in an uproar, and Sofie swooned, but not before proclaiming her fiancé's innocence. Jacobus went to court to testify on behalf of his friend.

The matter was resolved by an eyewitness account. Dr. Michael Maier, a philosopher, alchemist, and general favorite at court, happened to be passing through the royal gardens at two o'clock in the morning on the night in question, on his way to the Jewish cemetery for some specimens of human skull he needed for a particular experiment. Along his route he noticed someone dressed as a peasant and carrying a basket, going into the royal stables; this person was much shorter than Lieutenant Swoboda and walked with a limp. The vice chancellor put two and two together and freed Swoboda, who was then promoted to captain of the Royal Dragoons. He and Sofie were married, and Jacobus was made director of all of the imperial botanical gardens for his part in uncovering the plot. Malombra was arrested, made a full confession under torture, and was sentenced to sit in a small room in the tower of a castle, surrounded by food liberally laced with Aqua Toffnina.

The signature *Jacobus de Tepenec* appears on the first page of the Voynich manuscript, although it is now too faded to be seen, even with a magnifying glass. (It was discovered quite by accident when the photographer hired by Wilfrid Voynich to make copies of the manuscript spilled a developing solution on the page and the signature and inscription appeared.) Since Jacobus would not have used the designation "de Tepenec" until he had been raised to the nobility, this means that the manuscript did not come into his possession until after 1608.

An experienced botanist like Jacobus would have been the obvious choice to examine Rudolph's new possession. The many drawings of exotic, fantastic plants would have fallen into his area of expertise. If the plants could be identified, perhaps the names would provide a clue to the cipher.

But de Tepenec's signature implies ownership. It is unlikely that Rudolph would have voluntarily parted with so expensive a purchase, particularly one that, when decoded, promised to reveal the great secret of life. But in 1611, Rudolph was summarily deposed by the Bohemian Assembly. He had spent his declining years as a recluse, withdrawing further and further from his people and the everyday operations of government. He never left his castle and spent hours closeted with astrologers and alchemists. When he did rouse himself to rule, his actions were invariably unpopular. In 1609, he had tried to curtail the activities of his Protestant subjects and withdraw the hard-won privilege of toleration under which members of different sects were allowed to practice their various religions. Even Catholic Bohemians turned against him for this, and he was forced to reverse his decision and guarantee religious toleration instead. In April 1611, his people gave up on him entirely and crowned his brother Matthias emperor in his place. Matthias chose not to rule from Prague, as Rudolph had, but from Vienna. Rudolph died the following January and, with his brother absent in Austria, many of his treasures were looted. It is probably at this time that Jacobus took possession of the cipher manuscript, if only to protect it from thieves or damage.

There is no indication that Jacobus was able to identify any of the plants or break the code. Nor is it clear how long he owned this curiosity. Very soon, de Tepenec, rich and influential as he was, would be imprisoned and then banished from his estate in Melnick, just outside Prague. In the six short years following Rudolph's death, the political climate in Bohemia would become so combustible that the merest spark would ignite the entire region into flames. In 1619, all of Europe watched as one misguided prince awkwardly struck the match.

ALTHOUGH OFFICIALLY A HAPSBURG POSSESSION, and therefore nominally Catholic, the kingdom of Bohemia was actually an uneasy mix of Catholics, Lutherans, Calvinists, and Ultraquists, descendants of the fifteenth-century reformers (or heretics, depending on your point of view) John Hus and Jerome of Prague. No two groups worked together consistently, although in a pinch the Lutherans would side with the Catholics against the hated

Calvinists. At the time Matthias took over, each of these religions was still tolerated and allowed to maintain churches. Matthias tried to reimpose Catholicism as the principal faith, but he was too far away to be effective and had to rely on deputies. Besides, Matthias was old and childless. It was only a matter of time before a new emperor would have to be crowned.

In 1618, Matthias's rule was challenged by a dispute arising over whether the Protestants would be allowed to build a church in a primarily Catholic village. A meeting of deputy governors ended dramatically with Matthias's two Catholic representatives being tossed out the window of a castle. "We will see if your Mary can help you!" one of the Protestants called out after the bodies. Then: "By God, his Mary has helped!" as the victims landed safely in a big pile of refuse. The miracle notwithstanding, the Protestants took over Prague. The nearby township of Melnick, where Jacobus de Tepenec was district administrator, a heavily Ultraquist community, followed suit. They arrested the Catholic Tepenec and threw him into prison.

The very next year Matthias died, leaving open the question of succession. By hereditary right, the office of king of Bohemia should have gone to his cousin, the Archduke Ferdinand, ruler of Styria (in present-day Austria). But Ferdinand was virulently Catholic. The rebels knew that they could look for no mercy from that quarter, so they scouted around for a new candidate. They found him in Frederick V, the elector Palatine.

The Palatinate was a small, strategically placed triangle in Germany bordered by both the Rhine and the Danube, with its capital at Heidelberg. The head of this province was one of seven electors responsible for choosing the emperor. At the time of Matthias's death, Frederick was twenty-three years old and had been ruler of the Palatinate for ten years. Frederick was weak, provincial, inexperienced, and possessed of a conscience and scruples. He had recently married Elizabeth, daughter of James I of England, herself young, spirited, sheltered, and proud. When Frederick was offered the kingdom of Bohemia, Elizabeth is reported to have swayed him into accepting by opining that "she would rather eat sauerkraut with a king than roast meat with an Elector."

They seem to have been the only people in Europe (aside from their subjects) who did not recognize their unsuitability to govern a complex and volatile country like Bohemia. The Bohemians thought they were bringing in a rich, well-connected ruler capable of drawing on the resources of the English and the French to protect them from the Archduke Ferdinand and his ally, powerful, Catholic Spain. What they got was a pair of children throwing parties they couldn't afford. Frederick and Elizabeth had no knowledge of their adopted realm's traditions or customs. He scandalized the population by skinny-dipping; she was equally frowned upon for her too-fashionably-low-cut dresses and unwholesome reading material. With the most powerful states of Europe, Spain, France, and Germany, arrayed against them, they could marshal only one ally to their defense, and—this is not a joke—it was Transylvania.

They lasted one year. Then one of Frederick's fellow electors cut a deal with Ferdinand and invaded. The army moved swiftly toward Prague. Frederick's forces, under the command of his chancellor, staked out a defensive position on White Mountain, just outside the city. The chancellor assured the king that there was no need for alarm. Frederick and Elizabeth accordingly hosted a dinner that evening in the castle. The enemy attacked; the Bohemians were overrun. After dinner, Frederick thought he might just go outside and have a look at his troops. He met them, flying past, at the gates to the castle. The royal family had to evacuate so quickly that they almost left the baby behind. The plates were still on the table.

The end of Frederick's tenure marked the beginning of the Thirty Years' War. Over the next three decades, Bohemia would be systematically destroyed as one army after another attacked, conquered, looted, occupied, and was itself routed by a succeeding force. The capital city of Prague, in particular, as a hollow symbol of authority, became a focus of Catholic and Protestant ambitions. After the battle of White Mountain, the invading army closed off the city for a full week and treated itself to a wild looting spree that left the city's inhabitants, Catholic and Protestant alike, stripped of food and possessions. Afterward, Ferdinand claimed Prague as a Catholic stronghold, and the Jesuits poured in to undertake the task of reeducating the populace. Forty Protestant leaders were arrested, tried, and sentenced to a mass execution in the main marketplace. The Protestant citizenry were given the choice of conversion or exile; 150,000 chose the latter.

For Jacobus de Tepenec, the Archduke Ferdinand's victory over the hapless Frederick and Elizabeth meant a return to his home. A deeply religious man, Tepenec embraced the Jesuit cause. When he died in 1622 as a result of a riding accident, he left a fortune in gold and the contents of his Melnick estate to the Jesuit society. Although there is no specific mention of the book in his will, it is likely that the cipher manuscript found its way, with the rest of Tepenec's effects, to the Jesuit College, called the Clementium, at Prague.

For years afterward, the manuscript dropped from sight. The war continued, and each year the situation of the Bohemians deteriorated. In 1631, the king of Sweden banded together with one of the German electors and sent a Protestant army sweeping though Prague, but the victory was short-lived, and the Catholics, with the aid of the empire and Spain, were soon back in control. By 1639, when the Protestants came through again, Prague had been the scene of so much fighting that the city could no longer feed itself. Plague and poverty and death hung over the capital; John Baner, the Swedish marshal in command of the latest army, wrote: "I had not thought to find the kingdom of Bohemia so lean, wasted and spoiled, for between Prague and Vienna all is razed to the ground and hardly a living soul to be seen in the land." Already, a generation had been born that knew nothing but war. Yet this was the very year that the manuscript resurfaced in Prague. It seems incredible that a book that, after all, could not be read, could have survived such devastation. Perhaps it was saved by its unprepossessing appearance—it had neither brilliant hues nor gilt to make it recognizably valuable—and by the fact that it could not be eaten.

IN THE ARCHIVES OF THE PONTIFICIA UNIVERSITÀ GREGORIANA in Rome is filed a letter from Georg Baresch of Prague. This letter, dated 1639, is addressed to Anathasius Kircher, then professor of mathematics at the Roman College. Kircher, a Jesuit and renowned linguist, was the author of the recently published *Prodomus Coptus* or *The Coptic, or Egyptian Forerunner. The Coptic* was a primer in the language of early Egyptian Christians. Kircher's ability to read this archaic tongue had made him celebrated throughout Europe. It was only a matter of time, most scholars agreed, before he solved the puzzle of Egyptian hieroglyphics, which were themselves assumed to hold the key to uncovering the divine language that God had handed down to Adam and which had been lost to mankind with the expulsion from the Garden of Eden. This Adamic language—the very same tongue for which John Dee had been searching half a century before when he questioned his angels—was said to contain all of the great secrets of the universe, including the secret of life itself.

This is evidently the second letter of its kind that Baresch had written to Kircher—he makes reference to an earlier appeal for which he obtained no answer. The letter that survived was successfully delivered by the Reverend Father Moretus, a Jesuit priest traveling from Prague to Rome. In it, Baresch wrote:

> *Most Reverend Father,*
>
> *Expressing my highest regards, I wish that you may receive all happiness from Him who provides happiness . . .*
>
> *After the publication of the* Prodromus Coptus, *Your Reverence became famous in the whole world. In that work you had asked for help in finding additional material . . . from all those who might have something from which this work might be enriched . . .*
>
> *Since such a Sphinx in the form of writing in unknown characters was uselessly taking up space in my library, I thought I would not be unjustified in sending this enigma to be solved . . .*
>
> *Having thus transcribed (taking pains to imitate the writing) a certain part of this old book, which the carrier of this letter has seen with his own eyes and about which he can inform you, I send this writing . . . with the aim that (if Your Reverence would have been willing to undertake this investigation and convert these characters of unknown*

*creation to known letters) this toil could be of use either to [you] . . . or to me, or to the common good . . .*

*From the pictures of herbs, of which the number in the Codex is enormous, of various images, of stars and of other things which appear like chemical secrets, I conjecture that it is all of medical nature . . .*

*This work will be worthy of the effort of a virtuous genius, especially since this is not a work for all, which one may conclude from the fact that the author would hardly have gone to such lengths just to hide things which are open to the public . . .*

*I will be obliged to you for this, not just for what the work contains, but also all else that will become possible . . .*

*With this I recommend myself to Your Reverence and I wish you a happy, successful completion of this work . . . May the Almighty Lord preserve you for the community of literates.*

*M. GEORGIUS BARESCH*
*Prague, 27 April 1639, on the same day on which, in Rome in April 1605, I took up my studies at the University "La Sapienza" [The University of Rome].*

Very little is known of Baresch outside this letter. He seems to have been an alchemist, and something of a local scholar—his allusion to his days at the University of Rome was in the manner of one establishing credentials. If he went to school in 1605, he was probably born between 1580 and 1590, which means that he was old enough to have been infected by the scientific mania of Rudolph's court. He would have known of Tepenec. He had close ties to the Jesuits although was not a member of the order himself. How the manuscript came into his hands is a complete mystery.

There is no evidence that Kircher ever wrote back to Baresch, who continued to labor over the cipher manuscript throughout his life. It seems to have become an obsession with him—the light tone he adopted in his letter to Kircher about "a Sphinx uselessly taking up space in my library," masked a passionate desire to break the code. He never did. Perhaps, however, he broke the three-line key on the last page, and the notation that matched Newbold's attributing the manuscript to Roger Bacon which was labeled "1630" was his.

When Baresch died, he left his library, mostly alchemy books, to his good friend Johannes Marcus Marci of Kronland, a professor of medicine at the University of Prague. Marci had studied to be a Jesuit priest but decided instead to teach and practice medicine. He lectured at the university from 1630 until his death in 1667, eventually becoming dean of medicine and finally rector.

In 1638, Marci had gone to Rome and met Kircher, with whom he established a long and enthusiastic friendship, writing frequently, asking for books and advice. In 1640, Marci, too, had written to Kircher from Prague, recommending Georg Baresch and vouching for his credentials. In his letter he called Baresch *amicus meus* (my friend), and said that he came from an aristocratic family and was a scholar of intelligence.

When Marci came into possession of Baresch's legacy, he again wrote to Professor Kircher in Rome:

*Reverend and Distinguished Sir; Father in Christ:*

*This book, bequeathed to me by an intimate friend, I destined for you, my very dear Athanasius, as soon as it came into my possession, for I was convinced it could be read by no one except yourself.*

*Johannes Marcus Marci* BEINECKE RARE BOOK AND MANUSCRIPT LIBRARY, YALE UNIVERSITY

*The former owner of this book once asked your opinion by letter, copying and sending you a portion of the book from which he believed you would be able to read the remainder, but at that time refused to send the book itself. To its deciphering he devoted unflagging toil, as is apparent from attempts of his which I send you herewith, and he relinquished hope only with his life.*

*But his toil was in vain, for such Sphinxes as these obey no one but their master, Kircher.*

*Accept now this token, such as it is, and long overdue though it be, of my affection for you, and burst through its bars, if there are any, with your wonted success.*

*Dr. Raphael, tutor in the Bohemian language to Ferdinand III, King of Bohemia, told me the said book had belonged to the Emperor Rudolph and that he presented the bearer who brought him the book 600 ducats. He believed the author was Roger Bacon, the Englishman.*

*On this point I suspend judgment; it is your place to define for us what view we should take thereon, to whose favor and kindness I unreservedly commit myself and remain at the command of your Reverence,*

*JOHANNES MARCUS MARCI OF CRONLAND*

*Prague, 19th August, 1666 (5?)*

Enclosed with the letter was the cipher manuscript.

THAT A CURIOSITY LIKE THE CIPHER MANUSCRIPT should make its way to Anathasius Kircher was, perhaps, inevitable. Kircher was as celebrated for his wildly eccentric, exhaustive collection of oddities as he was for his Egyptian hieroglyphics. He had a *Musaeum* at the Collegio Romano, one of the first of its kind in Europe, jammed with everything from stalactites to a stuffed crocodile. He had traveled extensively, picking up specimens along the way, before finally settling down in Rome; also, as word of his collection grew, people—like Marci—sent him things. He loved gadgets, and whole rooms were devoted to the latest in seventeenth-century technology: magic lanterns (slide projectors), thermometers, clocks, megaphones (he linked these up from room to room to form an intercom system), spheres, *Smicroscopia* (Kircher's name for the microscope), magnets, trick mirrors. There were telescopes manned by Jesuits on the roof of the observatory tower, and, as a practical joke, one of the megaphones opened up behind a Greek statue in the foyer, which Kircher nicknamed the Delphic Oracle.

In the mind and spirit of Anathasius Kircher, mysticism, religion, and scientific curiosity held their last great collaboration. Born in eastern Germany on May 2, 1602, the feast day of his namesake, Kircher was the last of nine children. His father, Johann, was an educated man, a doctor of divinity who taught scripture at the local Benedictine monastery. Johann possessed a large private library and a questing, unorthodox mind, as evidenced by his decision to have his son, who attended Jesuit school, also trained in Hebrew by a rabbi.

Kircher would later write his life's story, *Vita admodum Reverend: P.A. Kircheri*, in which he detailed an eventful childhood. On one occasion, he went swimming in the mill pond with his friends and was swept up in the waterwheel; on another, he was pushed out onto the track in the middle of a horse race with the riders bearing down on him; still later, he caught a chilblain while skating, which turned septic, but he prayed ardently to the Virgin Mary and was cured overnight. These adventures did not end with puberty. During the Thirty Years' War he escaped an oncoming Protestant army by fleeing across the Rhine on an ice floe and narrowly escaped execution when he refused to shed his Jesuit robes and travel in lay attire. He tramped into the very heart of Mount Vesuvius and got lava on his sandals; he was robbed by pirates and shipwrecked in out-of-the-way places.

Because of these miraculous escapes, Kircher concluded that he had been singled out by God for great works. Naturally curious, he threw himself into scientific study. He acquired a telescope and observed sunspots. He dabbled in magnetism and optics. During his lifetime he would produce more than forty treatises on everything from the subterranean world to the cosmos.

His knowledge of antiquity allowed him to re-create Noah's Ark and the Tower of Babel in such mathematical detail that he

***Anathasius Kircher*** BEINECKE RARE BOOK AND MANUSCRIPT LIBRARY, YALE UNIVERSITY

was able to assert that the former measured precisely fifteen cubits and to debunk the theory that the latter was tall enough to reach to the moon. Based on reports from Jesuit missionaries in Asia, he wrote about Chinese calligraphy and exotic fruits like the pineapple, which he claimed could devour iron nails. He spent an enormous amount of time translating (incorrectly, as it turned out) and explaining the significance of Egyptian hieroglyphics. He was a student of John Dee's work and commented on it extensively. Whereas Dee had seen mathematics as the key to God and divine knowledge, Kircher believed light and magnetism to be evidence of divine truth. Also like Dee, Kircher was suspected of practicing the dark arts when the theatrical effects he had staged to entertain the archbishop of Mainz one evening—he also imitated flight with ropes and pulleys—turned out to be a little too realistic for his credulous audience.

A devout Jesuit, Kircher passed all of his accumulated knowledge through the filter of the teachings of St. Ignatius of Loyola. He continually searched for, and found evidence of, the Trinity in nature. All of the ancient cultures and religions he studied were also found to be in the Christian mold; scriptural significance was hidden in pagan symbols and allegories.

His talents were noticed by his superiors. Powerful Church patrons backed him, and he was awarded the chair in mathematics at the Collegio Romano. His arrival in 1635 could not have been timelier. The brilliant and popular Tuscan Galileo Galilei had recently been condemned to imprisonment by the Inquisition for publicly stating the Copernican notion that the earth went around the sun, and the Church desperately needed a champion with unimpeachable scientific credentials who would defend a strict scriptural interpretation of the heavens.

With Kircher, the Church made one last attempt to coopt science and subordinate it to faith. Kircher was a skilled showman. He dazzled the local peasantry with huge, dragon-shaped hot air balloons and beams of light that shot out of specially designed mirrors, and scared them silly with a magic lantern show about the flames of hell devouring a wicked soul. At Christmas, he set up a nativity scene complete with spotlights. But Kircher never questioned scriptural supremacy. Rather, he defended the Church's position: "Whosoever glorifies the dignity of the name of Christ, let him be especially careful not to philosophize beyond the limits of sacred canonical proscriptions," he wrote.

This was not hypocrisy. Kircher apparently resolved the centuries-old conflicting drives of faith and science by patching them together with the familiar John Dee form of spiritual mysticism. There is some evidence that he might have believed in the Copernican system but did not admit it publicly. Whenever he wanted to float an idea contrary to scripture, he said he had a dream. One could not be imprisoned or censured for dreams. As a result, "writing in support of a theological perspective that found great support among various Catholic rulers, Kircher enjoyed the privileged position of publicizing and elaborating a philosophy that was, in essence, the official position of the baroque church," the Stanford historian Paula Findlen concluded.

In fact, Kircher was the end of that line of mysticism that had embraced John Dee and the myth of Roger Bacon, which had had its heyday in the alchemy of the court of Emperor Rudolph. He was magnificent, flamboyant, spellbinding, and prolific. But he was also the Church's man, doomed to watch the great rolling ocean of scientific advancement wash over him, as a shell on the beach, before it finally ground him down to sand. For all the façade of mathematics and prolific knowledge, Kircher's science—when it was not outright fiction—was wishful, not empirical. The Royal Society could not confirm any of his results.

Isaac Newton, although much less well known, was already at work on his *Philosophia Naturalis Principia Mathematica* when Kircher died in 1680. Its publication seven years later sounded the death knell for those who still clung to the Bible as the literal source of all knowledge. The experimental science first formulated by Roger Bacon in his *Opus Majus* had finally come into its own.

AFTER KIRCHER'S DEATH, all of the curiosities in his *Musaeum*, carefully cataloged, became the property of the Jesuits in Rome. The cipher manuscript was never listed among his effects. Still, when Wilfrid Voynich pored through the chest containing the manuscripts of Pierre-Jean Beckx, 22nd General of the Jesuit Order, at Villa Mondragone 250 years later, there it was.

CHAPTER SEVENTEEN

# The Making of the Most Mysterious Manuscript in the World

. . .

WHEN WILLIAM ROMAINE NEWBOLD CONFIRMED Wilfrid Voynich's suspicion that the author of the cipher manuscript was Roger Bacon, it caused a sensation throughout the bibliographic and scientific communities. But Newbold had just gotten started. Two years later, on April 20, 1921, after enormous fanfare (generated largely by Voynich), Newbold presented a paper to an audience of top-level scientists and cryptanalysts at a meeting of the American Philosophical Society, where he announced his truly momentous achievement—he had cracked the primary code.

The results were astonishing. Although Newbold had been able to transcribe only about four percent of the cipher, it was apparent that this manuscript was perhaps the most important scientific document ever uncovered.

Within these encrypted pages lay knowledge centuries before its time, and in a fantastic array of scientific disciplines. According to Newbold's transcription, Roger Bacon had not only predicted the microscope and the telescope, but built them. Through his telescope, Bacon had viewed the Andromeda Nebula (illustrated in

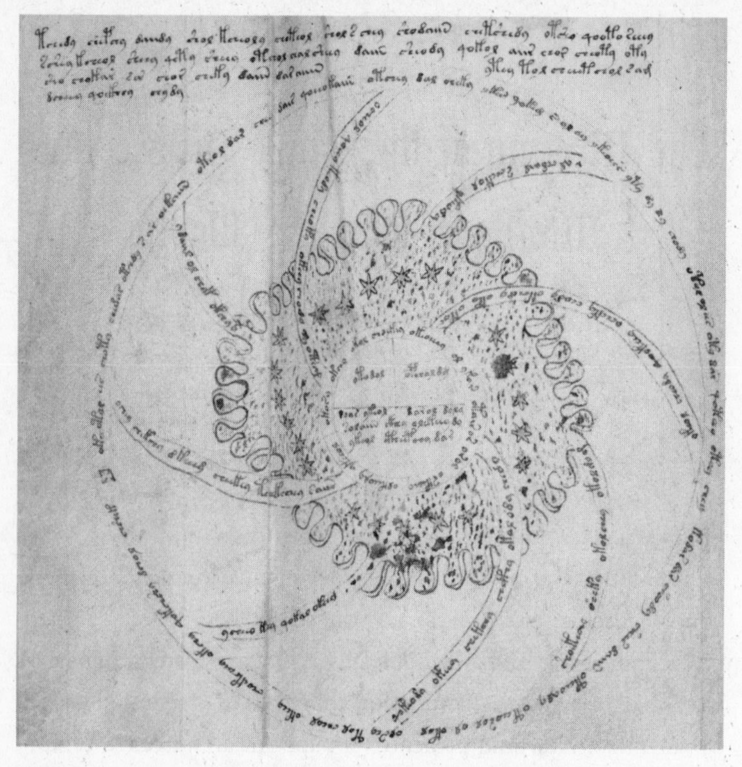

***Voynich manuscript illustration of the Andromeda Nebula as interpreted by Newbold*** BEINECKE RARE BOOK AND MANUSCRIPT LIBRARY, YALE UNIVERSITY

the manuscript), and, through the microscope, spermatozoa and a fertilized ovum (also with illustrations). He had predicted and described a solar eclipse in 1290, discussed steam power, dated a comet that appeared in 1273, described riots at Oxford between knights and ecclesiastics, also in 1273 (in which he played a decisive role, dispersing the crowd with an explosion of gunpowder), and detailed a process for refining copper ore.

Before he went into what was admittedly a circuitous method of encryption, Newbold assured his audience that he had no knowledge of chemistry and that he had been unaware of either the Andromeda Nebula or the Oxford riots. This, he claimed, was proof that he had not force-fit the transcription into a predetermined solution and that what he had found in the manuscript must be genuine.

The cipher itself was almost impossibly complex, six layers in all, although each of the layers had been described by Bacon in his letter of 1248. Newbold began with the Latin text on the last page, the first line of which read, "michiton oladabas multos te tccr cerc portas." Eliminating what he first assumed were nulls, Newbold came up with "michi dabas multos portas," which he translated as "thou hast given me many gates," although "multos" should have been spelled "multas." Aware that Bacon was well versed in the Kaballah, Newbold then assumed that "gates" referred to the convention in Kaballah that meant letter pairs, and so deduced that individual plaintext letters were represented by two adjacent letters in the ciphertext.

That got him closer, but he was still unable to make sense of the manuscript. Returning to the key and looking more closely at the individual letters, especially the *o* in "multos," Newbold noticed that what appeared to be an *o* written with flourishes was actually a symbol fashioned from three individual pen strokes. Each of these was highly reminiscent of a symbol in Greek shorthand with which Bacon had been familiar. Remembering that Bacon had praised shorthand as the best way of obscuring secrets, Newbold studied the shorthand, and patterns began to emerge.

Soon he had his alphabet—a substitution transcription of Latin letters, punctuation, and other symbols, based on frequency and position of the Voynich characters—but even then he was not yet home. Bacon, he claimed, had further hidden his true meaning

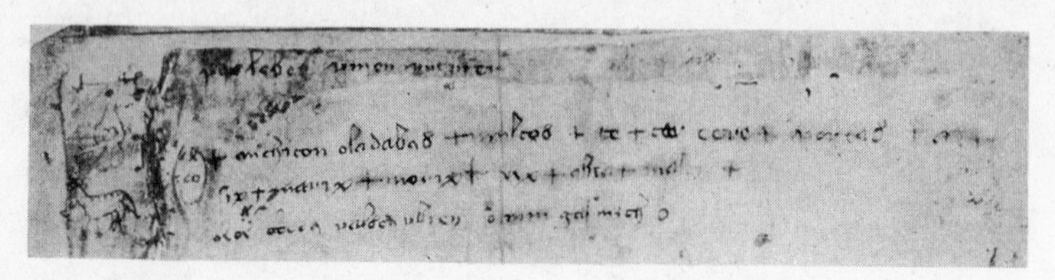

(1) michiton oladab*pqqoea* *nv*ultc*eqs* te tccr cerc port*qpipbpsetticlsd* m*tinq*
(2) six *m*arix mo*bq*ix v*nt*x ahta n*na* rict
(3) o*utloudpttkg* t*dttc*urq ualst*rtl* ubr en so nim gaf mith o

*Newbold's "key" worksheet reprinted in* **The Cipher of Roger Bacon** *by William R. Newbold*

in anagrams. The first confirmation occurred when he applied this alphabet to the discarded nulls in the key and came up with "R Baconi."

Had this pronouncement come from someone with lesser credentials than the Adam Seybert Professor of Intellectual and Moral Philosophy at the University of Pennsylvania, a man who had received only the second Ph.D. the university had ever granted in philosophy, a former dean of the graduate school, a scholar of soaring reputation and unimpeachable credentials—he was to be acknowledged the very next year by Theodore Roosevelt, Jr., for his code breaking for the navy—the members of the audience might have scoffed. As it was, none did.

Major newspapers and magazines across America reported the great discovery—the *New York Times* headline proclaimed, "Bacon 700 Years Ahead in Science." Both Newbold and Bacon burst from the cloistered confines of medieval scholarship into the popular arena and became instant celebrities. As David Kahn noted in

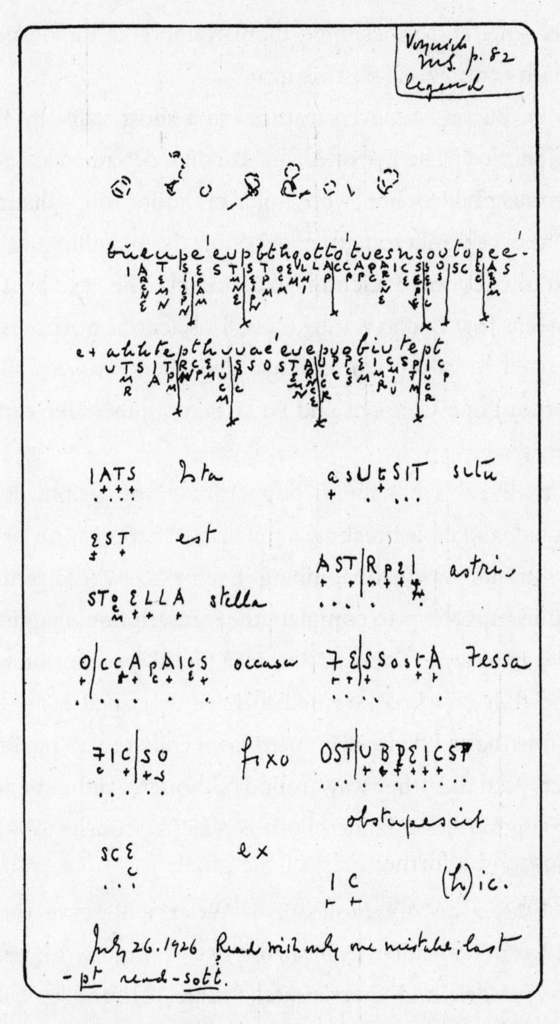


***Newbold's last completed worksheet reprinted in* The Cipher of Roger Bacon *by William R. Newbold***

*The Codebreakers*, "Sunday supplements had a field day." Even the Catholic Church jumped on board, willing to let bygones be bygones. Two separate articles in *Catholic World* lauded Bacon's achievements and proclaimed them testaments to the power of thirteenth-century scholasticism.

Later, Bacon was even featured in a short story by Rudyard Kipling entitled "The Eye of Allah." Bacon is described as "a learned and famous philosopher [who] holds his liquor too, valiantly," and is cast as a cantankerous friar decrying the unwillingness of the Church to encourage scientific experiment. The "eye of Allah" itself is a lens that Bacon wants to use for scientific purposes, but he is thwarted by more conservative and superstitious colleagues. The future Pope Clement and Peter Peregrinus make cameo appearances.

With Bacon's newfound popularity, cryptography itself became a fad, and code breaking a public fascination as never before. But the drama was just beginning. Even as Newbold returned to the manuscript to try to complete the transcription, nagging questions began to arise. *Scientific American* found the solution tortured and said that Newbold seemed to have described a one-way cipher—one that could be decrypted from code text to plaintext but not encrypted the other way around (although a different article in the same publication lauded both Newbold's conclusions and his scholarship).

In 1923, a history professor at Western Reserve University named Lynn Thorndike published a two-volume work entitled *A History of Magic and Experimental Science*. Thorndike had been working on this study for twenty years—he would add six more volumes, the last of which would be published in 1958—and it was so extraordinarily researched that he had unearthed documents in the Vatican Library that the librarians did not know they had. For all the detail, however—and Thorndike covered *everyone*—the longest single section was devoted to Roger Bacon, "who," he wrote, "in modern times has received so much attention and admiration at the expense of contemporaries and his age. Happily in the present volume we are in a better position to estimate him fairly." Then, caustically and with extremely selective use of data, he dismissed Roger Bacon as a minor figure in the history of science, vastly overrated and with no original work to his credit.

Debunking Bacon had become something of a crusade for Thorndike. Throughout the previous decade he had written articles in which he tried to diminish Bacon's achievements and paint him as a crank or mystic. What others had described as courage, Thorndike painted as stubbornness. Conviction became spite, persistence became jealousy. Thorndike's prejudice toward the Dominicans was palpable. He ridiculed Bacon for his belief in spirits and monsters but neglected to mention that Albert was every bit as superstitious, and Thomas even more so. Nowhere in Thorndike's work, for example, can one find Aquinas's conviction that wind storms were sent by a vexed God to discipline man. In *A History of Magic and Experimental Science,* Thorndike did not mention Newbold or the Voynich manuscript specifically but did include a scathing critique of Colonel Hime's assertion that Bacon had employed an Argyle cipher to disguise the formula for gunpowder.

Still, through the 1920s, Thorndike and *Scientific American* notwithstanding, most scholars continued to find Newbold's work brilliant and ingenious and his conclusions persuasive. Newbold won his greatest victory two months after his presentation when, in the July 1921 issue of *Harper's Magazine*, John Matthews Manly, who had been second in command of MI-8, wrote an article tentatively accepting both his methods and his conclusion. Manly had been one of the cryptanalysts who had failed to crack the cipher in the years immediately after Voynich acquired it. He had been a highly respected scholar before the war and afterward had returned to the University of Chicago to become chairman of the English department. Manly was the same age as Newbold (and Voynich) and had known and become friends with Newbold during the war.

The article also contained a compelling argument for the manuscript's authenticity. "It does not seem possible," Manly wrote, "to doubt that this is a real cipher, based upon a real language. No man in any century previous to the sixteenth could 'fake' a cipher system that possessed so many marks of genuineness as this and carry it consistently through nearly three hundred pages of closely written text." As to Newbold's esoteric assertion that the pen strokes in the key indicated shorthand, Manly added, "It would be easily possible to incorporate [Newbold's] microscopic signs into the strokes forming these symbols without in any way producing interference or difficulty."

With Manly on his side, Newbold had little to fear from the far less celebrated Thorndike.

As great a coup for Newbold as the *Harper's* article was, everyone in the field wondered whether Manly's old boss at MI-8, Herbert Osborne Yardley, who had also taken an interest in the manuscript but failed to crack the code, would weigh in as well.

Yardley, an inaugural member of the Hall of Honor at the National Security Agency's Cryptologic Museum, is one of the most famous and controversial figures in the history of cryptography, and the spur of perhaps the most famous epigram about the practice. He began his career in his early twenties as a code clerk in the State Department, but after privately cracking Woodrow Wilson's own enciphered telegrams, he was granted a commission in the army. After the United States entered the war in 1917, Yardley was sent to France as a cryptanalyst, and then, still not yet thirty, made head of MI-8, where he immediately recruited Manly to run the department in Washington. When Yardley returned home after the war, he induced the army to establish a permanent unit to penetrate and monitor other countries' coded messages, a group he floridly nicknamed "The Black Chamber," a French term for any government agency that opens and reads mail. Yardley's main objective was to track one particular nation's traffic, a country that he was convinced would soon be a threat to American security—Japan. Manly returned to the University of Chicago, but the men would remain good friends.

Yardley's group, whose funding came from the State Department as well as the army, was wildly successful, breaking the diplomatic codes not only of Japan, but of more than two dozen other nations as well, including Great Britain, France, and Italy. In 1929, the new secretary of state, Henry Stimson, upon discovering that America was spying on its allies, issued his famous (and astonishingly naive) pronouncement: "Gentlemen do not read each other's mail." MI-8's funding was cut off and Yardley summarily dismissed.

Yardley was bitter, furious, and out of a job. There weren't too many places that an unemployed spy could offer his services, at least without committing treason. With the Depression eating away at his meager savings, Yardley sat down and wrote his memoirs. Published in 1931, *The American Black Chamber* was a record of his wartime activities but also revealed the extent of the United States' cryptanalytic work in the 1920s. At the time, there was no law against publishing such a book, and it caused the pedictable red faces at the State Department. Stimson was livid. (As a direct result, two years later, a law authorizing hefty jail terms for anyone betraying America's cryptologic activities was passed.)

With no legal recourse against Yardley, Stimson wrote his own book, *The Far Eastern Crisis*, in which he strongly suggested, among other things, that the Japanese revise their codes. The Japanese took the advice (although Yardley's revelations themselves would certainly have been enough) and were therefore able to communicate in secret in the days leading up to Pearl Harbor. Stimson, of course, later served as Franklin D. Roosevelt's secretary of war during the conflict with Japan.

Yardley eventually worked for Chiang Kai-shek (and wrote another book, *The Chinese Black Chamber*) but was shunned by his native country until his death in 1958. He never took another shot at the Voynich manuscript.

MANLY'S CAREER TOOK A FAR DIFFERENT TURN. Although he had been Yardley's subordinate, he was more than twenty years older

***Herbert O. Yardley exhibit at the National Cryptologic Museum*** COURTESY OF THE NATIONAL SECURITY AGENCY

*John M. Manly (*left*), Edith Rickert, and an assistant returning to America in 1932 after a summer spent in Britain on the Chaucer project* UNIVERSITY OF CHICAGO LIBRARY

and lacked both Yardley's intuition and charisma. But he brought to any task the painstaking thoroughness of the accomplished scholar. Three years after publishing his article in *Harper's*, Manly began a project sufficiently massive to be worthy of Roger Bacon himself. His goal was no less than to complete an authoritative compilation of the works of Geoffrey Chaucer with the aim of producing the definitive text of *The Canterbury Tales*. He recruited Edith Rickert, a former colleague at MI-8, into the English department and then set to work on a project that would occupy both of them for the rest of their lives.

The Chaucer project is as fitting an example as one can find of the combination of intelligence, scholarship, doggedness, and endless immersion in detail that characterizes the elite cryptanalyst.

What Manly had in mind was as bibliographic as scholastic, and he set up an academic version of MI-8 to do the job. According to the University of Chicago, the task included "collecting, photographing, and collating all existing Chaucer manuscripts and studying their provenance. A Chaucer textual laboratory was organized . . . where a team of graduate students meticulously analyzed photostatic copies of the eighty-three fragments and complete manuscripts of the *Tales* found by Manly and Rickert. Lettering styles, paper markings, and types of ink were examined to find clues that might help establish each manuscript's origin."

Manly and Rickert stretched their personal finances to travel to Europe each year to visit museums, libraries, and private collections. They carefully examined manuscripts, searching for clues—any difference in the ink, paper, or binding might help differentiate a correct, early version of the text from a later imitation. Erasures were important, as was the manner in which the paper was trimmed. Provenance was vital, so they probed any surviving records concerning Chaucer's family or where he had lived for further clues.

The obsessive hunt for material began to wear down both Manly's and Rickert's health, as well as their finances. In fact, when Yardley came to Manly for a loan in early 1931, Manly was forced to turn his old friend down, after which Yardley immediately began his memoirs. Although Manly and Rickert completed what was to be the eight-volume *Text of the Canterbury Tales*, Rickert, exhausted, died in 1938, before the work was published. Manly died two years later, living just long enough to see his great effort in print. Both Manly and Rickert hold honored places in University of Chicago history, and there is now a John Matthews Manly distinguished service professorship at the university, which at one point was occupied by the great African-American historian John Hope Franklin.

One would think that while proceeding with such a single-minded effort even a scholar of Manly's capacity would have no time to revisit the Voynich manuscript.

But he did.

PROFESSOR NEWBOLD CONTINUED TO WORK and transcribe according to his alphabet and anagrams, uncovering more and more of the marvelous legacy of Roger Bacon. Then, on the evening of September 25, 1926, he was struck with what the *New York Times* described as "acute indigestion." The next day he was dead. His friend and colleague Roland Grubb Kent collated all of Newbold's transcriptions, notes, and papers on the Voynich manuscript and published them in 1928 as *The Cipher of Roger Bacon*.

Bacon remained largely unchallenged as a great mind of science who was centuries before his time. Thorndike continued to grumble—he wrote in 1929 in *American Historical Review,* "There is hardly one chance in fifty that Roger Bacon had any connection to the production of the Voynich manuscript"—but the fascination of the cipher was irresistible, and naysayers were dismissed with a wave of sour grapes.

Wilfrid Voynich died in March 1930. In his will, he named a panel of five experts, led by his "associate" Anne Nill, to sell the manuscript to an appropriate—and wealthy—public institution. (There is no specific evidence of how far this association went, but Voynich and Nill spent lots of time together, even taking sea voyages on which his wife, Ethel, was not present. Recently, a rumor has surfaced that Ms. Nill was the Voynichs' adopted daughter, but, as with almost everything else having to do with this manuscript, nothing is certain.)

Another member of the panel was Manly. That he stipulated that each of the five receive $3,000 (in 1930 dollars) for their time and effort is an indication of what Voynich expected the manuscript to sell for. Newbold's widow was to receive ten percent of the proceeds, Roland Grubb Kent another five percent. The remainder Ethel and Anne Nill would split sixty-forty.

Before it was actually sold, however, one of the five experts did something that was to vastly affect the selling price and the ultimate destination of the manuscript. In 1931, a forty-seven-page article appeared in *Speculum,* a journal published by the Mediaeval Academy of America. Its title was "Roger Bacon and the Voynich MS," and the author was John Matthews Manly.

In the article, Manly revealed that Newbold had been sending him progress reports until his death, and, he wrote, "I told Professor Newbold of my conclusions and gave my reasons for them in several letters." After Newbold's death, he had hoped to "let the subject rest," but too many members of the scientific community had accepted the transcription and Manly felt forced to reexamine the issue in "the interests of scientific truth." He did not mention in the article that a university, private library, or government agency was about to spend possibly a million dollars for it.

Then, with obvious sadness and reluctance, Manly proceeded to demolish the work of his friend, brick by brick, argument by argument. "In my opinion," he began, "the Newbold claims are entirely baseless and should be definitely and absolutely rejected."

Manly's objections were so fundamental that after reading his article, those who had accepted Newbold's claims could not believe they had done so. Anagramming, for one thing, was so subjective that even Newbold himself had come up with three distinct decipherments from the same passage. (The letters l-i-v-e, for example, can yield five distinct words, and the number of possibilities increases exponentially as more letters are added to the base.*)

There were other obvious errors. In one case, Newbold copied the ciphertext incorrectly and then came up with a deciphered version anyway; in another he came up with a Baconian transcription from a passage that had not been part of the original manuscript but had been added later. As for the "shorthand," the seeming separation of letters into strokes was merely a result of uneven paper texture and the ink line breaking over the centuries. Some of the history was wrong as well. According to Manly, an authority on the period, there were no "knights" at Oxford in 1273, and therefore the account of the riots was obviously wrong.

But what of Newbold's claim that he knew nothing of any of these phenomena before transcribing the cipher? For this, Manly had saved his most devastating assessment. Newbold *must* have known, Manly concluded, even if he was unaware of his knowledge, and therefore "his decipherments were not discoveries of secrets hidden by Roger Bacon but the products of his own intense enthusiasm and his learned and ingenious subconsciousness."

Manly tried to lessen the blow in the end by stating that for his dedication, perseverance, and unwillingness to admit that the manuscript could not be read, Newbold's "record of defeat was

* They are *live, evil, vile, veil,* and *Levi.*

none the less a record of scholastic heroism." From that day forward, however, no one thought William Romaine Newbold a hero. In fact, in his chapter in *The Codebreakers* entitled "The Pathology of Cryptology," David Kahn's opening lines are about Newbold.

Although he had been reduced to poster boy for wishful cryptanalysis, Newbold was not its only victim. In 1926, Manly himself had produced a work entitled "Some New Light on Chaucer," an attempt to link the pilgrims in the *Canterbury Tales* to real people. That work was greeted with the same enthusiasm as was Newbold's, and it turned out to be just as spurious.

It was in the same year as Manly's article that Albertus Magnus was made a saint, and everything that Newbold, those who had supported his version, and Voynich himself had said was now discarded. One could almost hear Lynn Thorndike sniffing, "I told you so." The market for the manuscript dried up—a sale could probably not even have recouped the experts' fee—so Ethel Voynich ended up sticking it in a safe-deposit box, where it would remain for the next thirty years, until her death in 1960 at age ninety-six. (Further muddying an already unclear association, Ethel Voynich left her sixty percent ownership of the manuscript to Anne Nill, who had become her inseparable friend after Voynich's death.)

Through it all, however, there was no getting around that Marci had believed Bacon to be the author and so had Kircher and Rudolph. As to Dee, he had either believed the manuscript to be Bacon's work or known that it wasn't—perhaps he had never seen it at all. The possibility of forgery or hoax now became one of the more active hypotheses. People started looking at its provenance a little more carefully, and suddenly nothing about the manuscript was certain.

There were even some doubts that the manuscript was ever in Dee's possession. Dee liked to write his name in his books or jot down notes in the margins. None of these notations was present. Still, in the upper-left-hand corners of the pages were numbers written in a handwriting that has been confirmed by experts to be Dee's own. (There are dissenters even here, however.) This and his son's later, unrelated reminiscence that his father had been puzzling over a treatise written entirely in "Hieroglyphicks" while in Prague, and Dee's coincidentally obtaining six hundred ducats from Pucci, would all point in the direction that Dee had in fact sold the cipher manuscript to Rudolph.

The next question became, if Dee did own it, where did he get it? It has been suggested that it was one of the manuscripts he uncovered while scouting the English countryside, or was perhaps a gift from his former employer the Duke of Northumberland, himself a beneficiary of the spoils of the monasteries under Henry VIII and his son, Edward VI. But if this was so, why was the book not listed in the catalog that Dee composed for his library in 1583? Why is there no mention of it in his diary? Dee very much liked to show off his treasures to Walsingham and Elizabeth. Surely a manuscript written entirely in cipher would have been of interest to the queen or the head of her secret service. And if he had owned the manuscript since the days of Queen Mary, why take it abroad with him in the 1580s? Or, did he pick up the manuscript while abroad, either in Crakow with Count Laski or in Prague itself, as something that had been brought to his attention as a curiosity?

Then there was the possibility of fraud. While nothing in Dee's long career would indicate that he himself would knowingly be party to a swindle, if either Edward Kelley or Francesco Pucci was involved in the transaction, one cannot rule out that the manuscript was an extremely clever forgery, written as part of a deliberate scheme to capitalize on Rudolph's credulity and the topicality of Roger Bacon in Bohemia. The illustrations of plants and little naked women in the manuscript, both particularly to Rudolph's taste, lent credence to the idea of a hoax.

The fraud or forgery hypothesis remained popular until the next serious effort at decipherment was undertaken, this time by two other inaugural enshrinees at the Hall of Honor at the Cryptologic Museum of the National Security Agency—William and Elizebeth Friedman.*

The Friedmans, generally considered the greatest cryptanalysts who ever lived, came to the field during World War I when both were working for an autocratic, self-promoting, eccentric millionaire named George Fabyan. Fabyan, who liked to be called "colonel" although the title was honorary, was the scion of a Boston textile family who relocated to Illinois, bought Riverbank, a 350-acre estate, which he had remodeled by Frank Lloyd Wright, and proceeded to there create the world's first think tank. His projects ranged from the intensely practical—acoustic plaster was developed at Riverbank—to the absurdly fanciful, such as an attempt

* The hoax hypothesis has recently been rekindled by Gordon Rugg, a British computer scientist. He claims that an examination of pattern and frequency of symbols reveals that the text might have been created by Edward Kelley, using one of his favorite enciphering techniques. Kelley, says Rugg, would have filled out a grid 40 by 39 with the ideographic symbols, then overlaid it with another piece of paper with three diagonal squares cut out. By moving the top grille over the bottom grid, a ciphertext can be created. Even if true, this would not account for the detail in the illustrations and the length and intricacy of the manuscript. Kelley seems sure to have lacked the patience for such an undertaking.

to develop an antigravity machine. During his lifetime, Fabyan was perhaps best known for persuading a Chicago judge to rule that Francis Bacon was the true author of Shakespeare's plays.

Today, there is plaque at the Riverbank Laboratories, presented by the National Security Agency, which reads, "To the memory of George Fabyan from a grateful government." What the government is grateful for is a department of codes and ciphers—run by the Friedmans—that produced some of the most significant advances in cryptologic science ever recorded.

Neither Friedman arrived at Riverbank to work on codes. William, a Russian immigrant whose original given name was Wolfe, was a graduate of Cornell who was recruited to run some stock-breeding experiments in Fabyan's genetics department. Friedman had become a geneticist only because genetics was then associated with agriculture, and an immigrant's son could get a free undergraduate and graduate education if majoring in some aspect of farming.

Elizebeth Smith (the spelling was to keep her from being called "Eliza"), a Quaker's daughter and Shakespearean scholar from the Midwest, had been hired as an assistant to Mary Wells Gallup. Gallup had convinced Fabyan that Francis Bacon, using the biliteral cipher, had hidden tidbits of juicy Elizabethan gossip (not the least of which was that he was the queen's illegitimate son) within the text of Shakespeare's plays, of which, of course, he was the author. Another revelation buried in the biliteral cipher was of the levitation machine, which Fabyan attempted to build following the simple step-by-step directions. It failed to lift anything off the ground, however, including itself.

Fabyan spent a fortune to have Francis Bacon's manuscripts and early editions of Shakespeare's plays shipped to Riverbank to expose additional wonders. Examining the typefaces on individual letters in all these manuscripts—which were in fact printed in any number of fonts—was a hefty task, so Friedman (who also helped make photographic enlargements) and Smith were recruited to help.

Although they didn't find any evidence of steganographics, Friedman and Smith did find each other. In 1917, the twenty-six-year-old Friedman and the twenty-five-year-old Smith were married, and the most famous partnership in the history of cryptology was born. Almost immediately, Friedman showed a flair for the work that approached genius.

Moving to Riverbank's cipher department, which by then was doing some government work, Friedman cracked the "unbreakable" British field code in a matter of hours, then went on to break German codes and even a book cipher used by Hindu militants who were trying to use the war to gain Indian independence from Britain. While at Riverbank, Friedman produced a series of papers that essentially recast the science of cryptology—a word Friedman himself invented. Eventually, he joined the army and trained most of the cryptanalysts—another word he invented—who worked for Yardley at MI-8. The most important change, however, according to David Kahn, was that it was William Friedman who first linked cryptology to mathematics, thereby creating a science out of an art.

Elizebeth Friedman enjoyed enormous success in the field as well, mostly in breaking codes used by drug smugglers and rum runners during Prohibition, and was later to set up a secure communications network for the International Monetary Fund. She was also instrumental in breaking a Japanese spy ring during

World War II, and is famous for the dismissive "Our office doesn't make 'em, we only break 'em," said to a salesman who had come to sell her department a new coding device.

In late 1944, after a quarter century of brilliant success and achievement, the Friedmans turned to the Voynich manuscript. (Through Manly's work, William Friedman was already aware of the intricacies of the problem.) In the middle of the war, they set up a study group of overworked cryptologists to work after hours to see what they could make of the document.

The Friedmans worked from a remarkable set of documents produced by a teacher of language and history at the Catholic University of America in Washington, D.C., named Theodore C. Peterson. Father Peterson, the first of many obsessed amateurs who would take a shot at the manuscript, was the assistant to Professor Eugene Xavier Henri Hyvernat, an expert in Near Eastern languages. Professor Hyvernat had worked with Voynich on issues of provenance at the Pierrepont Morgan Library in New York, and, after Voynich's death in 1930, Ethel Voynich had sent him her husband's photostats, asking if Hyvernat could make some sense of them. Hyvernat was in poor health and could participate only sporadically (although he had gone to Rome a couple of years earlier at Voynich's request to check Kircher's correspondence), but Father Peterson became hooked.

He made a second photostat of every page and eventually began the slow process of creating a handwritten copy of the entire manuscript. As he worked, he left notes in the margins of each page as to what he thought a particular illustration or passage might represent. The copy was not completed until 1944. Father Peterson also undertook a study of concordances, a listing of each word, the pages on which it appeared, and the words immediately preceding and following. Despite all his efforts, he had made no progress toward a solution. Father Peterson continued to work on the manuscript until his death in 1966, but he lent his notes and photostats to the Friedmans later in 1944.

An interesting sidelight emerged from Father Peterson's work, however. A botanist and Benedictine friar, Hugh O'Neill, was consulted on the plant illustrations and claimed to identify two as plainly originating in America. One of the two was to O'Neill obviously a sunflower, the seeds of which were brought to Europe by Columbus after his first voyage, thus dating the manuscript after 1493. Although some other botanists have concurred, cryptanalysts have pointed out that in a manuscript in which so many of the drawings were fanciful, the resemblance might have been completely coincidental, or it may not have been a sunflower at all.

After the Friedmans were given the material, they decided that the first task was to conduct a frequency analysis of the various symbols in the cipher. In pre-scanner days, this was an extraordinarily dreary chore. The Voynich manuscript guru James Reeds, who has a Ph.D. in statistics and was formerly a member of the mathematics of communication department at Bell Labs, noted in a 1994 article in the scholarly journal *Cryptologia,* "One of the difficulties facing anyone attempting to read the Voynich manuscript is the tedium of preparing a transcription into conventional alphabetic or numerical symbols, in order to make possible frequency counts and concordances. Such a transcription requires study of often unclear photocopies [all the Friedmans had], and, if carried out with conscientious regard for accuracy, is very time consuming." Dr. Reeds, who has devoted more than a decade to the manuscript, added, "My experience suggests that after transcribing about ten pages the would-be VMS reader loses interest in the task, starts worrying about his eyesight and stops work, leaving more than two hundred pages untranscribed."

Still, not only did Friedman's study group transcribe the entire manuscript, assigning letter values and number values to the various symbols, but they transferred the entire package to IBM punch cards and produced a printout of their work. Dr. Reeds noted that this "might well be the first example of a machine readable edition of a text prepared for scholarly purposes."

Friedman and his group's method for approaching the task provides a revealing look into the cryptanalyst's craft. Although computers have taken over compiling and provide an almost unimaginable sophistication of both quantitative and qualitative analysis, the methodology of the cryptanalyst—part trial and error, part intuition, part mathematics—has remained largely the same.

Since the group probably began expecting a cipher, they started with techniques appropriate to unraveling that form of steganographics. According to Dr. Reeds, "In 1944 this kind of work was done with punch card equipment, so it was natural for them to put the Voynich manuscript text onto punch cards. So some scheme for rendering the Voynich manuscript characters into punch card symbols had to be used. But before they could start the main work of transcribing, the [study group] had to pick a transcription alphabet."

This means that Friedman and his colleagues had to first properly identify individual symbols and then assign each a letter or number to proceed with the analysis. When dealing with a series of unknown symbols, such as those in the Voynich manuscript, this first task particularly becomes a purely subjective exercise fraught with peril.

As Dr. Reeds points out, "The kinds of mistakes that the wrong choice leads to can be imagined by supposing a future race of beings trying to decode our writing system. If they mistakenly assume that 'm' and 'n' are the same letter (because they don't believe the exact number of humps could be important) or that 'h' and 'n' are the same letter (because they differ only in length of a single stroke), or that 'n' and 'u' are the same (because they are rotated versions of each other) their analysis will be made harder. On the other hand, if they think that 'm' and '*m*' are genuinely different letters, or that 'A' is fundamentally different from 'a,' their analysis might become bogged down with irrelevant minutiae."

Even if a cryptanalyst gets the symbols right, there is always the question of nulls, symbols that merely key other symbols or combinations, the insertion of nonsense phrases, and biliteral phrases (Francis Bacon's method of inserting messages under perfectly intelligible text).

Trying to peer through the murk, Friedman's group produced reams of worksheets and grids to give some sense of character frequency and placement. (All this was being done, it should be remembered, by men and women already fully engaged in one of the most vital and time-consuming tasks of the war effort. Friedman himself had collapsed from overwork in January 1944 and had convalesced in Walter Reed Hospital for three months with what was diagnosed as "psychoneurosis.")

The group began with frequency counts of letters and letter combinations, indexes, and concordances. After a good deal of

TENTATIVE LIST OF CHARACTERS[1]

| |  | | |  | |
|---|---|---|---|---|---|
| A | 1. | P | O | 15. | 4 |
| B | 2. | F | P | 16. | E |
| C | 3. | H | Q | 17. | C |
| D | 4. | D | | ~~18.~~ | |
| E | 5. | G | R | 18 ~~19.~~ | T |
| F | 6. | A | S | 19 ~~20.~~ | S |
| G | 7. | R | T | 20 ~~21.~~ | I |
| H | 8. | K | U | 23 ~~22.~~ | HZ |
| I | 9. | 2 | V | 21 ~~23.~~ | PZ |
| J | 10. | O | W | 24. | DZ |
| K | 11. | L | X | 22 ~~25.~~ | FZ |
| L | 12. | N | Y | 25 ~~26.~~ | V |
| M | 13. | M | Z | 26 ~~27.~~ | Y |
| N | 14. | | | ~~28. (space)~~ | |

27. space
28. par
29. illegible character

[1] As agreed at Meeting on 9 June 44

Transcribed by W.F.F. 13 June 44

***Tentative list of Voynich characters as assigned by William Friedman*** WILLIAM F. FRIEDMAN COLLECTION, GEORGE C. MARSHALL FOUNDATION

wrangling and innumerable corrections, they came up with a tentative list of twenty-nine characters, of which nineteen were letters (most of the Latin alphabet), three were numbers, four were

two-letter combinations, two were format instructions (space and paragraph), and one was unknown.

Their transcription alphabet looked like this:

VOYNICH MANUSCRIPT
ALPHABET FOR TRANSCRIPTION

|  |  |  |  |
|---|---|---|---|
| 1. | P | 16. | E |
| 2. | F | 17. | C |
| 3. | H | 18. | T |
| 4. | D | 19. | S |
| 5. | G | 20. | I |
| 6. | A | 21. | PZ |
| 7. | R | 22. | FZ |
| 8. | K | 23. | HZ |
| 9. | 2 | 24. | DZ |
| 10. | O | 25. | V |
| 11. | L | 26. | Y |
| 12. | N | 27. , | space |
| 13. | M | 28. ,, | paragraph |
| 14. | 8 | 29. ø | illegible charac. |
| 15. | 4 | | |

AUTHORIZED BY Mark Rhoads

**_Friedman transcription_** WILLIAM F. FRIEDMAN COLLECTION, GEORGE C. MARSHALL FOUNDATION

"They probably picked the scheme they did because of visual resemblance of the manuscript characters to various letters and numbers," Reeds noted. "They must have thought that the gallows characters [21–24] without pedestals paired up with those that had pedestals, and so thought of using a following Z as a kind of accent mark."

Then, after they produced the punch cards, got their printout, and were prepared to plunge into the actual decipherment, World War II ended. With the war over, funding for Friedman's department dried up and the group was disbanded before an exhaustive statistical profile could be developed. Still, just from the work they had done, one or two points caught Friedman's eye.

Every language has "entropy," a measure of disorder that is derived from the degree of repetition of words or letters. For simple languages entropy is very low, and for more complex languages it is somewhat higher. If the manuscript was a hoax—nonsense made to look like ciphertext—entropy would be higher still, a good deal higher than in genuine language patterns. What Friedman found was just the opposite. Entropy in the Voynich manuscript was much lower than that for most real languages—so low, in fact, that it was later determined that the only actual language with so much repetition was Polynesian. Yet the repetition was never longer than five-word segments, unlike the long repeated phrases that one would expect in a hoax. Also, in most cases, it is words that repeat, rather than phrases, the manner in which genuine language is structured.

Last, and perhaps most important, was that, as Manly had first pointed out, the manuscript was so specifically rendered, so consistent in structure, the illustrations so detailed and finely drawn, that if it was a hoax, it was sufficiently elaborate to have taken years to fashion, even if the text was meaningless. If Kelley had fabricated the manuscript to hoodwink Rudolph, he would not have needed to go to nearly so much trouble.

William Friedman rejected the hoax alternative. He also rejected the notion that the manuscript could be written in substitution cipher, no matter how sophisticated. Instead, he came up with a different theory entirely, one that has guided most of the research ever since. When asked to explain his theory in an article for *Philological Quarterly,* Friedman chose to do so in anagram cipher, the solution to which he sealed in a time-stamped envelope, inviting readers to try and unravel it. He wrote:

> *I put no trust in anagrammatic acrostic cyphers for they are of little real value—a waste—and may prove nothing.—Finis.*

To prove Friedman's point, at ninety-six letters, this anagram lends itself to a variety of plausible solutions. Readers of *Philological Quarterly* sent in a number of possibilities, including, "To arrive at a solution of the Voynich Manuscript, try these general tactics: a song, a punt, and a prayer. William F. Friedman." Another was, "This is a trap, not a trot. Actually, I can see no apt way of unraveling the rare Voynich Manuscript. For me, defeat is grim." When the envelope was opened, however, the actual message was

> *The Voynich MSS was an early attempt to construct an artificial or universal language of the a priori type. Friedman.*

What Friedman had in mind was pasigraphy, a universal language based on classification in which symbols stand in for words. Efforts to create such languages were thought to have begun in the sixteenth century but had their origins in the Orient, where ideographs represented entire concepts rather than simply words or phrases. Francis Bacon had attempted such a system, as had Leibniz. No modern scholar has ever attributed such an effort to Roger

***William Friedman*** COURTESY OF THE NATIONAL SECURITY AGENCY

Bacon, but given his interest in the study of comparative language and accuracy of translation in scholarly work, a universal language, particularly one with such obvious applications to scientific study, would have been an ideal creation. Also, if Francis Bacon was attempting to formulate an artificial universal language, it is not at all outrageous to speculate on whether the idea might have originated elsewhere. Finally, when John Dee's library was sacked, many of his books and papers were stolen. If Roger Bacon had hypothesized about a universal language, evidence of it might well have disappeared.

In the Friedman pasigraphical variation, broad categories of knowledge were given symbol or "letter" equivalents and placed at the beginning of a word, then basic subcategories, also assigned a symbol (from the same "alphabet"), were then placed in the second position, and so on. If this was the method used, sets of letters would repeat frequently at the beginnings of "words" but

*Elizebeth Friedman* COURTESY OF THE NATIONAL SECURITY AGENCY

would then break off into variances as the words got longer. This appeared to be the case with the Voynich manuscript, but Friedman's attempts to study the matter further were hampered by a shortage of both time and money. (By this time, the manuscript had been so discredited that when Friedman tried to obtain funding from a private foundation he was denied on the grounds that the manuscript held nothing new.)

In the 1950s, the Friedmans returned to the other Bacon to try to determine once and for all whether it was possible that he was the author of Shakespeare's plays. They were both by now in their sixties and, while still active, had to choose projects for which there was more than an outside chance of resolution. In 1955, they published *The Shakespeare Ciphers Examined*, in which they demolished not only Mary Wells Gallup's reading of biliteral messages—Gallup, it seemed, although honestly believing her readings, had assigned typefaces based more on what she wanted to find than on similarities or differences—but also a number of other Baconian theories that were a good deal quirkier. In most of these cases, the Friedmans found not that the various interpretations were impossible, but simply—like Newbold's Voynich transcription—that they were subjective and ambiguous, thereby violating the one rule of cryptology that conspiracy theorists most often ignore—the need for a secret message to have one and only one solution.

WHILE FATHER PETERSON AND THE FRIEDMANS WERE ANALYZING the data, two "solutions" were put forth, with great fanfare. In 1943, a Rochester, New York, lawyer named Joseph Martin Feely self-published a book entitled *Roger Bacon's Cipher: The Right Key Found*. Feely, who became interested in the cipher after reading Newbold and Kent's book, hired a team of researchers, who then completed an exhaustive frequency analysis of both the manuscript and several of Roger Bacon's original texts. From these results, Feely attempted a substitution based on the most common letters and symbols. Frustrated both by Bacon's penchant for abbreviation and the differences in classical and medieval Latin (to say nothing of never seeing the original manuscript and working completely from the illustrations in the book), Feely all the same attempted to "crib" the words that were associated with the drawings, then move from that to a detailed transcription. (A "crib" in cryptology is a piece of ciphertext with a plaintext equivalent that is known or can be deduced without decryption.) He concluded that the manuscript was a scientific journal with a specific emphasis on gynecology. (The naked women were actually ova, and he found references to ovaries and tubes throughout the work.) As to the cipher itself, Feely concluded that it was a simple substitution of abbreviated Latin, most likely written by Roger Bacon.

No one either then or since has found his solution persuasive.

A more compelling resolution was put forth by Dr. Leonell Clarence Strong, a distinguished geneticist and cancer researcher from Yale and (of course) amateur cryptographer. Claiming that he could decipher anything, and also working from the illustrations in the Newbold and Kent book, Dr. Strong asserted in a 1945 article in the journal *Science* that the manuscript was written in a "peculiar double system of arithmetical progressions of a multiple alphabet." He refused, however, to publish the details of his method, claiming that such a disclosure might be used by the nation's enemies.

Dr. Strong's solution "revealed" that the manuscript was actually an herbal and sex manual written by Anthony Ascham, a sixteenth-century alchemist and astrologer and the brother of Roger Ascham, the humanist tutor to Edward VI and Elizabeth I. (He also claimed to have decoded Ascham's name on the very same page that held O'Neill's sunflower.) The text contained, according to a contemporary newspaper account, "an extremely candid discussion of women's ailments and practical matters of the conjugal bed," as well as a recipe for an herbal contraceptive. Ordinarily, such a bizarre theory would have been greeted with a snort, but Dr. Strong had a thirty-year record of brilliant work in medical research. On the other hand, after Newbold, academic credentials did not carry the weight they once might have. Elizebeth Friedman observed, "He said little about [his method], but what he did say made no sense to cryptologists."

. . .

IN THE YEARS AFTER WORLD WAR II, interest in the manuscript gradually waned, and, when Ethel Voynich died in 1960, Anne Nill, now the sole owner, decided to sell it. Rather than the millions that Voynich had expected when he wrote out his will, its highest bid came from a book dealer, Hans Kraus, who paid $24,500. He listed the item in his catalog at $160,000, then tried to find a buyer at $100,000. Although the Friedmans had set up a second study group in 1962, interest in the manuscript had fallen off, and the group soon disbanded. Unable to get even the lower price for "the most mysterious manuscript in the world," in 1969, Kraus finally donated it to the Beinecke Rare Book and Manuscript Library at Yale, getting nothing but a tax write-off. The Beinecke librarians filed the manuscript as MS 408 and expected, along with Kraus, that that would be that. The Voynich manuscript would become just another oddity—a unique but essentially anonymous item, blending in to what is doubtless one of the finest bibliographic collections in the world.

It did not work out quite that way.

CHAPTER EIGHTEEN

# MS 408

. . .

MS 408 IS ANYTHING BUT ANONYMOUS. In fact, it is by far the most requested item in the entire Beinecke collection. The Voynich manuscript has its own curator, who is assigned to field requests and queries; there is a Voynich "kit" sent to the curious; and images of various Voynich pages are posted regularly on the Beinecke Web site. One of the librarians noted facetiously, "We get busloads of people from Cleveland pulling up to the curb and coming in wanting to see the Voynich manuscript." If auctioned today, the manuscript would almost certainly, after almost three quarters of a century, finally garner the kind of money that Voynich expected.

INTEREST IN THE MANUSCRIPT WAS REKINDLED soon after it arrived at Yale. A member of the university's philosophy department and former army intelligence officer named Robert Brumbaugh had become fascinated with the manuscript thirty years before, when, as a young student of medieval philosophy, he had read both Newbold's account and Manly's reply. He found himself "unable to resist" taking a look now that the manuscript was at the Beinecke.

Soon, Brumbaugh was visiting the library in every spare moment to pore over the symbols and the illustrations, and he even recruited his son to help. While examining the first page under ultraviolet light, Brumbaugh's son confirmed that there was indeed a transcription of the key attributing the manuscript to Roger Bacon, dated 1*30, which Brumbaugh concluded must be 1630, although he initially thought the illegible second number was lower than a six.

Brumbaugh himself focused on the key on the last page, as well as those illustrations that he could identify as representing genuine medicinal plants. (He admitted that "there were not many.") Using a different selection of letters and applying a completely different methodology than had Newbold, Brumbaugh nonetheless also arrived at the conclusion that Roger Bacon's name was embedded within the key. He later wrote, "The line [in the key] that opens: MICHI CON . . . BA. The CON . . . BA could be a Bacon anagram (like the modern pig-Latin ACONBAY) but one would not be sure without further confirmation. The letters OLADA between CON and BA are not a standard Latin word. But, in standard medieval cipher (where each letter of ciphertext stands for the letter three places later in the standard alphabet) OLADA deciphers as RODGD." The trailing isolated *O* at the end of the third line of the key, he added, would also have a plaintext equivalent of *R*, which added some confirmation.

From there, Brumbaugh worked furiously to try to make some sense of the ciphertext that encaptioned the medicinal plant section. He noticed that some of the characters appeared to be similar to a Renaissance Milanese code that he had run across, which had a base that was in numbers that transcribed into letters. This time, employing his daughter as well as his son, Brumbaugh came up with a partial decipherment based on a numbering system so astoundingly complex that it made Newbold's method seem straightforward. His plaintext transcription yielded what seemed to be part mystical herbal and part astrological manual.

However, even if part of the key did decipher into "RODG.R BACON"—questionable enough in itself—Brumbaugh hastened to add that it did not necessarily mean that Bacon himself had penned the manuscript. Citing bits of evidence found in the text and illustrations (for example, Sagittarius is drawn wearing a sixteenth-century cap) and other more circumstantial data (the lettering in a set of Leonardo da Vinci manuscripts "matched many characters" in the Voynich cipher), Brumbaugh decided that the manuscript must have been written much later than the thirteenth century. Because the "key" was so easily deciphered, he further observed, it must have been a plant, part of an elaborate fraud concocted by someone, probably Dee or Kelley, who knew that evoking the name of the famed friar would elicit great interest and render the document more valuable—possibly far more valuable. And no one, Brumbaugh concluded, would have valued a mysterious Roger Bacon cipher manuscript more than the gullible and willing Rudolph. Although the text was not gibberish and could be read—and probably would be someday—the content would hardly qualify it as a great work of science.

Brumbaugh's explanation attracted a good deal of interest, but not one professional cryptologist signed on to his solution. Their disagreement seemed to be as much with his conspiracy theory as the details of his solution. No one, they decided, would go to the enormous and unnecessary trouble of constructing a genuine underlying text to fool someone who could never have deciphered it in the first place.

While Brumbaugh was working at Yale, interest had also been renewed at the National Security Agency. Working at the agency at the time, on loan from the United Kingdom, was Brigadier John Hessel Tiltman, the man, according to General Communications Headquarters (GCHQ), the English version of the NSA, "generally acknowledged as the greatest British code breaker of all time." Tiltman was a prodigy. Born in 1894, he entered Oxford at age thirteen. He joined the army at the beginning of World War I, and after the war was assigned to a newly formed code-breaking unit called Government Code & Cipher School. He founded the military section of GC&CS in 1930 and during the years that followed established himself as a premier cryptanalyst by breaking most of the codes used by Communist parties around the world in communicating with Moscow. When tensions rose with Japan at the end of the 1930s, Tiltman set himself to preparing his country for war. He broke the main Japanese naval code in 1939, then, realizing that there was a paucity of cryptologists in his section sufficiently proficient in Japanese, taught himself enough of the language in six months to allow him to break the codes used by Japanese military attachés.

When GC&CS moved to Bletchley Park in 1939 (where the tortured mathematical genius Alan Turing would break the Nazis' seemingly impenetrable Enigma code), Tiltman remained in charge of the military section. While he was there, he attacked the cipher with which the German High Command communicated by radio. Given the code name "Fish," the cipher had been created by the Lorenz Company in response to the High Command's request for a secure means to send their radio messages. Lorenz had built a machine with letter wheels and keys similar to Enigma, but which transmitted signals in a thirty-two-character set of symbols called the Baudot code, not in the twenty-six-letter alphabet as did the Enigma machine. Without ever seeing the Lorenz machine, beginning entirely with cribs, Tiltman and his group eventually broke Fish and were able to intercept German military communication at the highest level.

After the war, in 1950, William Friedman sent Tiltman copies of the last twenty pages of the Voynich manuscript—a section with only text and no illustrations—as well as Newbold's book. Planning to use Tiltman as a "control," Friedman asked him what he could make of the symbols.

Tiltman first performed a quick frequency analysis, focusing on the symbols that appeared most often in the text and how they combined with other symbols. He discovered that some symbols appeared with overwhelming incidence at the beginning of words, some in the middle, and others at the end. What's more, there were symbol groups that seemed to cluster at the beginning of many words. Tiltman concluded that these were roots and that the ending symbols were suffixes, and therefore Friedman's hypothesis of a pasigraphic system was most likely correct. If this was true, of course, a substitution cipher was not a possibility.

He remained fascinated by the manuscript and, while he was in Washington in the late 1960s, began an investigation of some early attempts at artificial language, but there was no parallel system that had been created early enough or with sufficient similarity to provide any clues. He also undertook a history of herbals, but once more he could find nothing that would be of help. No herbal or medicinal manuscript before 1700 bore any resemblance to the document he was studying.

By 1970, Tiltman had become yet another of the greatest cryptanalysts in history who had reached a dead end in trying to decipher the Voynich manuscript. Like Friedman and Manly before him, he had unearthed new clues but found no solution. Tiltman's labors in many ways epitomized the entire frustrating history of the efforts to decrypt the manuscript. Cryptology involves an almost superhuman degree of patience and persistence, to say nothing of an artist's eye. After trying cribs, substitutions, concordances, frequencies, combinations, after checking provenance, medical history, philology, botany, and astronomy, after subjecting the cipher to both computer analysis and human instinct, no one was able to (persuasively) penetrate the code. There was too much detail in the manuscript, too much order, too much precision, for it to be a fraud, but too much obscurity to provide an answer.

While Tiltman was working on the cipher, another veteran NSA analyst, Prescott Currier, noticed that, particularly in the herbal sections, some of the pages were written tightly, the letters small and cramped, while in most of the text the script was looser and more flowing. Studying the characters separately, Captain Currier theorized that not only were there two styles in the manuscript but two separate languages, which he named "Voynich A" and "Voynich B." He couldn't penetrate either language, but, if he was correct, Currier had eliminated a potential source of error. While not everyone has agreed with Currier's assessment, the dual-language hypothesis has been incorporated into most of the subsequent research.

In the early 1970s, Tiltman, now nearing eighty years old,

heard that another young analyst at NSA named Mary D'Imperio had taken an interest in the manuscript. He turned over coordination of the Voynich transcription effort to her. (Whether the analysts were engaged in their Voynich activities in their spare time or as part of an official effort to see if the Voynich cipher had any utility has never been made clear.) In 1972, D'Imperio chaired a seminar, and then, four years later, published a monograph—actually under NSA auspices—entitled "The Voynich Manuscript: An Elegant Enigma." Although there is little in this work that is new, it is extraordinarily detailed, complete, and well researched, covering even the most peripheral aspects of the manuscript's scholarship. If nothing else, it provides remarkable insight into the methodology of contemporary cryptanalysis and supplies some pretty revealing hints as to how they go about things at the NSA. A quarter century later, this work is still considered definitive, but D'Imperio personally (if she was in fact a real person and not an NSA phantom) has faded back into the NSA mist, never to be publicly seen or heard from again.

THROUGH THE LATE 1970S AND 1980S, with nothing more heard from NSA analysts (at least for attribution), the spotlight was left to maverick amateurs. One of the most inventive entries came in 1978 when a self-styled philologist named John Stojko produced a solution in which he claimed that the manuscript was a copy of correspondence written in Ukrainian. In a book he called *Letters to the God's Eye*, Stojko claimed that before encryption all the vowels had been removed from the plaintext. The consonants were then encrypted by use of a secret alphabet with Etruscan roots. As a result, Stojko—who is Ukrainian himself and claimed to be able to read Etruscan—came up with sentences such as "Mr. is that you who believe in your believe? Miss, your Sust will renew eye and will rule. Tell me Eye of God; will God's eye of Sust heal?"

Stojko's solution quickly went the way of Feely's. Not only did the text of the letters bear no relation to illustrations on the pages on which they appeared, but in his transcription he had inserted vowels and word spaces arbitrarily. In addition, Stojko, who also claims to be a historian, was said by others interested in the manuscript to often deviate from "the generally accepted history of the Ukraine."

There was little limit to the lengths that amateur cryptologists would go to be able to claim that they had cracked the code. In 1987, Leo Levitov, a New Jersey physician, produced a detailed solution in which he claimed that the Voynich manuscript was actually the last surviving liturgical document of the great Cathar heresy of thirteenth-century France and that the Cathars themselves were an offshoot of a cult that worshipped the Egyptian goddess Isis. Dr. Levitov noted,

> *There is not a single so-called botanical illustration that does not contain some Cathari symbol or Isis symbol. The astrological drawings are likewise easy to deal with. The innumerable stars are representative of the stars in Isis' mantle. The reason it has been so difficult to decipher the Voynich Manuscript is that it is not encrypted at all, but merely written in a special script, and is an adaptation of a polyglot oral tongue into a literary language which would be understandable to people who did not understand Latin and to whom this language could be read. Specifically, a highly polyglot form of medieval Flemish with a large number of Old French and Old High German loan words.*

Further, according to Dr. Levitov, the streams of naked women depicted a Cathar sacrament called *endura,* the cutting of a vein in order to bleed to death in a warm bath.

Like Newbold, Brumbaugh, and even Stojko, Dr. Levitov, after years of work, had created a tightly knit logical structure and transcription alphabet to support his case. Although his solution was at first met with some tentatively favorable comments, his methods and conclusions soon came under more harsh scrutiny. His transcribed text did not match details in surviving Cathari records (the existence of which he was seemingly unaware), and the *endura* was almost always a ritual fast at the end of life and not a form of suicide. Dr. Levitov's linguistics was also called into question.

In 1991, in order to promote more serious research, Jim Gillogly, a computer scientist at the Rand Corporation, the California think tank, and Jim Reeds at AT&T Laboratories used the fledgling Internet to form a loosely knit colloquium of other Voynich devotees around the world. Some, like Reeds and Gillogly, had worked with encryption techniques professionally, whereas others were merely dedicated amateurs. The group soon grew into a cyber-study network with members from the United States, Europe, and Asia. "It's very orderly," observed Dr. Reeds. "Everyone is listened to politely, even the crackpots."

Today, while there are dozens, perhaps hundreds, of Web sites that present one theory or another as to the manuscript's origins, the Reeds-Gillogly network (although both of them would deny that the network is "theirs") continues to dominate the field. With its formation, study of the actual characters has drifted beyond the layman's grasp into the rarefied world of higher mathematics, statistics, and computer science. This is sad in a way, since, historically, it has been the Levitovs, Stojkos, Feelys, and Strongs who have provided most of the fun.

And technology has done no better. A number of alphabets and transcriptions have been proposed, but despite subjecting the manuscript to the most sophisticated tests—Zipf's Law, for example—current researchers are not really any closer to a solution.* Someone even completed a master's thesis in library science with a statistical analysis that demonstrated that the "gallows characters" in the manuscript were not nulls. Jim Reeds noted, "No real progress has been made in reading the Voynich manuscript, except in the purely negative sense of shooting down screwball theories."

Any progress that has been made is in tracing provenance—Baresch and Beckx have recently been added to the list of owners—and fleshing out the sometimes lurid details of Wilfrid Voynich's life and that of his family. But none of this has aided decryption at all. Most of the experts and professional cryptologists who have studied the manuscript are convinced that it is not a hoax, but even that is only supposition. The absence of any attempt to scientifically date the paper or ink—Yale refuses to do either—renders the general conclusion that the manuscript dates from the fifteenth or sixteenth century rather than the thirteenth also merely a guess.†

* Named after Harvard linguistic professor George Kingsley Zipf, this is a tool to determine the degree of repetition of words or symbols and therefore to measure entropy in language. Strictly stated, it is "the frequency of occurrence of some event (P)—such as words or symbols—as a function of the rank (i) when the rank is determined by the above frequency of occurrence, is a power-law function $P_i \sim 1/i^a$ with the exponent a close to unity." Here again, mathematical precision has only served to confirm what Friedman saw by instinct.

† Curators at the Beinecke claim that even if it was discovered that the manuscript was written on thirteenth-century paper, it does not necessarily mean that it was written by a thirteenth-century author. Anyone wishing to make the manuscript seem to have been completed in the thirteenth century might have used older paper to help perpetuate the illusion. If, on the other hand, the paper is sixteenth-century, it could have simply been a copy of an earlier work. Ink dating, they say, is not sufficiently reliable to be definitive.

. . .

CRYPTOLOGY HAS COME A LONG WAY since Bacon wrote his "Seven Ways of Concealing Secrets" in 1248, transformed, as most other forms of communication, by the vast capacity of computers. The art—or science—is now almost entirely computer based, and encrypted messages are a regular part of the life of anyone who uses the Internet. Interestingly, technology has not made cryptology more complex but, at least in its basic approach, more simple. Cryptographics is no longer keyed by mysterious symbols, page templates with cutouts, or Kaballist numerology—these are all considered just grist for the supercomputer's mill—but instead in complex algorithms or numbers so large that only computers can manipulate them.

Still, in some ways, there has been no real progress at all. From Roger Bacon's time up through the Friedmans', any code or cipher fashioned by human ingenuity was susceptible to decipherment by the same means. No matter how brilliant the mind that fashioned a code, an equally brilliant mind might break it. In the technological age, we have simply transferred that same principle to machines—no matter how powerful a computer is thought to be or how impenetrable an encryption, another more powerful computer might always come along and decrypt it.

The search for the unbreakable code has led cryptology to the horizons of particle physics and a system called quantum cryptography. In quantum cryptography, a key to a computer encryption, changed at rate of up to 1,000 times per second, is transmitted over a fiber-optic link in a stream of photons, indivisible particles of light that have no mass but carry electromagnetism, one of the fundamental forces of nature. Photons, of course, travel at the speed of light. In quantum cryptography, these photons are polarized (the same technique is used in sunglasses, laptop screens, and liquid crystal displays), then sent in two streams. Because the electromagnetic energy in the system can be precisely measured, anyone trying to eavesdrop or break into the stream will cause an immediately detectable fluctuation that will disrupt the stream and render the message meaningless. When the intended receiver cannot decode the message, he or she will realize that someone has interfered.

Quantum cryptography is, in theory, unbreakable because the very act of trying to break the code destroys it. If history has taught us anything, however, it is that there is a technology just over the horizon that will trump the technology on the horizon. And, of course, even with supercomputers, giant streams of numbers, and polarized photon cryptology, the centuries-old Voynich manuscript remains unread.

With all the failures and dead ends, it becomes tempting to wonder if Newbold might have been correct back in 1921. More than eighty years later, no one has really done any better. And there are indeed those who insist that Newbold was unfairly attacked by his detractors and his version is largely correct, which would also mean that Roger Bacon is the manuscript's author after all.

In a 1997 article in the New Agey *Journal of Borderland Research,* a computer cryptographer named Aldrich Daimler argued that when people like the Friedmans turned the art of cryptography into a mathematically based science, they may have made one type of cryptanalysis more efficient, but they also effectively closed off an important area of inquiry.

According to Daimler, a cryptology consisting almost entirely of statistical analysis, complex algorithms, and the generation of immense streams of numbers completely ignores what in the Middle Ages was actually a creative art. Medieval cryptographers such as Roger Bacon *intentionally* employed steganography in a complex manner, rife with ideographs, anagrams, obscure and varying keys, puns, biliteral ciphers, and anything else they could think of, often in the same text. It is perfectly reasonable, therefore, Daimler argues, that Roger Bacon would employ all seven methods for hiding secrets in one document. What is certainly true is that Roger Bacon had a sense of humor. It was in his nature to have fun with his ciphers. He wrote them as much to playfully challenge his friends as he did to hide information. The tortuousness of Newbold's solution, then, rather than detracting from its veracity, actually reinforces it, and the Voynich manuscript becomes a prime example of a thirteenth-century cryptographer's art.

In this, Daimler may well be correct. Even Jim Reeds admits that the Friedman study groups, since they were so mathematically oriented, never considered the illustrations as being integral to the cipher instead of merely objects of it.

Daimler is most certainly writing from the fringe. Borderland Sciences was founded "for the purpose of investigating into realms normally beyond the range of basic human perception and physical measurement." According to its Web site, Borderland currently specializes in alternative medicine, radionics, psychotronics, water technology, dowsing and radiesthesia, Tesla technology, and new energy. Daimler himself, according to Borderlands, seems to have disappeared. Regardless of the source, however, the article contains some persuasive specifics.

He noted, for example, that while Manly claimed that the "pen strokes" in the *o* in *multos* in the key were due to the ink cracking, the ink in many other manuscripts of the same period had held up just fine. In another section, he takes issue with the most serious charge in Manly's article, that anagramming is incurably subjective and that Newbold strayed all over the text to force-fit the solutions to anagrams into a predetermined transcription. But even if Daimler is correct and Newbold never went further than a few characters in any direction, there is no saying that Newbold came up with the correct meaning.

Other than the spectral Daimler, there are few today who are willing to say that Roger Bacon had anything to do with the Voynich manuscript at all. Bacon may well not have been involved, but, with all the conjecture and inconsistency in every other theory, Newbold's hypothesis makes the most sense. If the key does decrypt into Bacon's name—either by Newbold's reasoning or Brumbaugh's—and the probability that Edward Kelley would go to all that needless effort to fool Rudolph is small, why not accept Newbold's decryption, at least in part? None of the artificial language arguments has gotten anyone anywhere. Newbold's solution at least matched the illustrations, and he may *not* have subconsciously known of the Andromeda Nebula or riots at Oxford.

But whether or not Roger Bacon had any impact on the Voynich manuscript, the Voynich manuscript had a very definite impact on Roger Bacon.

CHAPTER NINETEEN

# The Unfinished Legacy of Roger Bacon

· · ·

INSTEAD OF BEING THE MAKING OF BACON'S REPUTATION, the cipher manuscript turned out to be the ruin of it.

Before Voynich and Newbold, Roger Bacon's scientific work had been undergoing a slow but solid renaissance. In 1861, Emile Charles had published *Roger Bacon, sa Vie, ses Ouvrages, et ses Doctrines*. This detailed, highly researched four-hundred-page biography was the most ambitious work on Bacon ever undertaken. Without making any outrageous claims as to Bacon's achievements, Charles lauded him as a martyr to the advancement of knowledge. Two years before, J. S. Brewer, a professor at King's College, London, had edited the *Opus Minus, Opus Tertium, Compendium Studii Philosophii*, and the 1248 letter and included a long introduction that came to much the same conclusion as Charles's work.

The Brewer edition and the Charles biography sparked a number of articles and studies that denoted Bacon as a serious student of science. Then, in 1896, Andrew Dickson White published *A History of the Warfare Between Science and Theology in Christendom.*

***Andrew Dickson White*** CORNELL UNIVERSITY

White, trained as a historian, had been a New York state senator, diplomat, and cofounder and first president of Cornell University. He was also a noted bibliophile, and, in 1891, he donated the 30,000 books in his collection to the university. His medievaliana still represents the cornerstone of Cornell's collection.

Although a devout Christian (he was Episcopalian), White detested dogma, and Cornell had been established as a nonsectarian university, to the fury of many conservative religious leaders. Accusations of godlessness were not about to make Andrew Dickson White back down, however. In his two-volume work (every bit as well researched as Thorndike's), White traced scientific discipline after scientific discipline—chemistry, geology, astronomy, botany, physics—showing the progression from science based solely in scripture through the period when empiricism began to tip the scales to when scientific inquiry could finally proceed unencumbered by dogma. *A History of the Warfare Between Science and Theol-*

*ogy in Christendom* caused an enormous sensation when it was published and is today, more than a century later, still in print.

White saved his harshest condemnations for ecclesiastic zealots, such as Thomas and even Albert, who had "been drawn or driven from the paths of science into the dark, tortuous paths of theology," and his highest praise for men who not only were great scientists, but also showed moral courage in maintaining their belief in the truth against the repression of theocrats. For this latter category, his epitome was Roger Bacon. White wrote, "His *method* of investigation was even greater than its *results*. In an age when theological subtlizing was alone enough to give the title of scholar, he insisted on *real* reasoning and the aid of natural science by mathematics; in an age when experimenting was sure to cost a man his reputation, and was likely to cost him his life, he insisted on experimenting and braved all the risks. Few greater men have lived."

Interest in Roger Bacon increased. The next year, an English translation of the complete *Opus Majus* was published, followed by many of Bacon's Oxford lectures and earlier works. In 1914, to mark the seven hundredth anniversary of his birth, a group of international scholars formed a committee (Colonel Hime of the gunpowder article was vice chairman) to produce a series of new authoritative essays detailing all aspects of Bacon's achievements.

So, when Thorndike first published his articles, he was attacking a far less sensational, more solidly based Roger Bacon than that which emerged after Newbold, and so his smugness, sarcasm, and obvious bias did not win many over.

After Manly's 1931 *Speculum* article, that all changed. Scholars, embarrassed at having been duped, fled from the fanciful theories of the discredited Newbold and rushed to the comfortable academic meticulousness of Thorndike, tending to overlook that in his subjectivity and fussiness Thorndike was just as unreliable. Detail without context became the touchstone of Bacon scholarship, and the pendulum swung, as pendula will, past dead center and on toward the other extreme. The lingering need to debunk Bacon, to demonstrate that he wasn't the great scientific figure that he had been made out to be, has dominated the research ever since.

As late as 1997, Jeremiah Hackett, in the introduction to his otherwise excellent *Roger Bacon and the Sciences: Commemorative Essays,* wrote that Bacon "is known to modern readers in the profile which was constructed by popular writers on science in the 19th century . . . The image of Bacon as the Romantic hero of science who six hundred years before the Modern Scientific Revolution had anticipated that revolution continued into popular imagination in the 20th century."

But to which modern readers is Hackett referring? Bacon ceased being a "Romantic hero" to the general public seventy years before. As for the academic community, the only serious study of Bacon's life since Thorndike was Stewart Easton's *Roger Bacon and His Search for a Universal Science.* Easton, while assuring the reader that he had "tried to keep from any bias for or against" Bacon, went on to say, "I have worked on the assumption that he cannot have been unique, and that his originality, as, indeed, all human originality, has rested on his treatment of materials familiar to large numbers of people in his time." Would Easton have taken the same approach if he were writing about Newton or Einstein? As it was, through the filter of that assumption, Easton came to the conclusion, not surprisingly, that little that Bacon did was original (or unique).

There has been a good deal of recent scholarly work on specific areas of Bacon's studies, such as *perspectiva*, geography, mathematics, and experimentation. The goal in each case has been to determine whether Bacon's achievements qualify as science ahead of its time. Rarely does it stand up to this test.

As it stands, since Manly disemboweled Newbold in 1931, no one has been willing to say that Roger Bacon was prophetic, not because of any of his particular theories or experiments but simply because of the manner in which he viewed scientific inquiry. Lacking this designation, Bacon has now been relegated to the second tier—bright, perhaps, but ultimately of little importance in relation to the progression of human thought.

IN TODAY'S WORLD OF MODERN CONVENIENCES, of cell phones and pagers, stealth fighters, nuclear submarines, electron microscopes and radio telescopes, it is sometimes easy to forget that the ability to pursue science unimpeded, to let imagination wander where inquiry leads, to investigate and reach beyond ourselves, is not an entitlement but a right that was fought for, as were other rights fought for, sometimes to the death. This struggle has shaped our history and our souls. In it are the seeds that define us as a species. It was curiosity, not Eve, that tempted Adam.

Justifiably, we reserve our highest esteem for those who championed the way, who braved the crevice that snaked through superstition and fear on one side and repression and orthodoxy on the other. This task required not just intelligence but the commitment and moral courage that served as inspiration for others.

Thomas Aquinas was an honorable, pious, utterly sincere man. There is not a shred of evidence that he ever acted out of malice or spite or for personal advancement. There is no record of his showing anger or behaving with anything but charity toward his enemies. He worked solely for the glory of a God and a religion that he believed in deeply and fully. And, unlike many who claim to speak for God, Aquinas lived the values that he demanded of others. Even his adversaries acknowledged his personal goodness. There can be no more comprehensive a definition for a saint.

Roger Bacon, while also honorable, pious, and sincere, had a far less congenial personality. He was capable of temper, supreme impatience, resentment, and professional malice. He most surely limited his effectiveness by his unwillingness to moderate either his views or his approach.

Still, for all that, Thomas Aquinas narrowed the boundaries of curiosity—the very essence of human progress—and Roger Bacon sought to expand them. His failure to convince others of his century who were not yet ready to follow in no way lessens his achievement.

Roger Bacon is one of the most significant and irreplaceable figures in the history of science. His approach to knowledge and experiment is now the fundamental basis of scientific inquiry. Whether or not he actually produced the incredible inventions with which he is sometimes credited is unimportant. What he undoubtedly did produce was the method for creating them.

# ACKNOWLEDGMENTS

NAVIGATING THE LABYRINTH of material surrounding the Voynich manuscript was a formidable task. We are indebted to Stephen Jones, John Monahan, and Christa Sammons at the Beinecke Rare Book and Manuscript Library at Yale for answering questions, supplying technical assistance and photographs, and allowing us access to the manuscript itself. On the code side, our perspective was enlarged by Dr. James Reeds, who was unfailingly generous in his willingness to supply explanations of some of the more arcane facets of the decryption efforts. Rafal Prinke and René Zandbergen were also very helpful, as was David Kahn, who helped point us in the right direction. The public affairs office at the National Security Agency was (to our surprise) very willing to help, although in the spirit of the agency we are not going to mention anyone by name.

Oxford and the Bodleian Library were obviously very important stops on our journey, and we are indebted to Ross King for arranging for us to be allowed access to John Dee's legacy. When we got to Duke Humfrey's Library, all the materials we had requested were waiting for us thanks to Martin Kauffmann. We would not have both been able to study the manuscripts if Sara Langdon, a secretary at the admissions office, had not been so generous as to watch our daughter for us and even take her out to tea.

On the home front, Rachel Rice at the Westport Public Library was extraordinarily helpful and good-natured in her efforts to supply us with the multitude of interlibrary loan books we requested as background. At Doubleday, Gerry Howard supplied his usual insight, wit, and enthusiasm to the editing process, which helped us to improve the book immeasurably, Rakesh Satyal always responded immediately to any request we might have, and Alison Ker Miller did a super job copyediting. As always, our daughter, Emily, inspired and reassured us, which must have been difficult, as we were often the ones acting like children.

Our agent, Jed Mattes, died of pancreatic cancer in July 2003. This book would not have been possible without him. Jed was that most extraordinary of men who could operate on an exalted plane both personally and professionally. We miss him terribly. His partner, Fred Morris, held the agency together with grace and strength during Jed's illness and beyond, and we are indebted to him for the support he gave us during an extremely trying time.

## BIBLIOGRAPHIC NOTES

• • •

ANY EXAMINATION OF THE LIFE AND WORK of Roger Bacon necessarily begins with a study of the man's own words. This is no hardship, as Bacon was a lucid and passionate writer, and his work makes for compelling reading. Happily, many of Bacon's manuscripts have been ably translated into English from the original Latin, so we encourage anyone wishing to go further into this subject to seek out copies of his major works. The definitive *Opus Majus* is the two-volume 1928 edition, translated by Robert Belle Burke. The *Opus Tertium* and *Compendia Studii Philosophii*, edited by James S. Brewer, with extensive commentary, is available in Latin in an outstanding 1859 edition. *Roger Bacon's Letter Concerning the Marvelous Power of Art and Nature and Concerning the Nullity of Magic* is also available in translation by Tenney L. Davis. David S. Lindberg has recently published the first-rate *Roger Bacon's Philosophy of Nature: A Critical Edition of De multiplicatione specierum and De speculis comburentibus.*

The body of secondary literature, as may be expected, separates pretty evenly between early admiration and later skepticism, with Lynn Thorndike's formidable but deeply flawed series *A History of Magic and Experimental Science* as the dividing line. Of the earlier works, A. G. Little's 1914 *Commemorative Essays* and John Henry Bridges's *The Life and Work of Roger Bacon: An Introduction to the Opus Majus* are the most useful. But there is no shortage of contemporary scholarship focusing on Bacon's life and work. Lindberg's work and *Roger Bacon & the Sciences: Commemorative Essays* edited by Jeremiah Hackett are as scrupulous renditions of Bacon's work as have ever been produced. Unfortunately, each of these scholars, particularly Hackett, by restricting themselves to a study of Bacon's specific achievements, repeats Thorndike's error of missing that the ultimate value of Bacon's work was in approach and point of view.

By far the best work on Grosseteste, Bacon, and the Oxford school is the incisive and eminently readable *Robert Grosseteste and the Origins of Experimental Science, 1100–1700* by A. C. Crombie. What sets Crombie apart from other contemporary Bacon researchers, such as Hackett, is evenhanded treatment. Richard W. Southern's *Robert Grosseteste: The Growth of an English Mind in Medieval Europe* is another excellent discussion of Grosseteste's intellectual odyssey and his effect on everyone that followed.

While virtually all of St. Thomas's immense volume of work is readily available, there is no worthwhile critical biography of Aquinas. The work that is considered definitive, *Friar Thomas D'Aquino,* was written by James Weisheipl, himself a Dominican, who accepts even the stories of Thomas's visions without a trace of skepticism. (Jeremiah Hackett is a Weisheipl protégé.) Equally, Thomist logic is almost never subjected to critical analysis and dissection as was Bacon's. Ralph McInerny, a professor at Notre Dame University (and the author of the Father Dowling mysteries), has written extensively on both St. Thomas's life and works but, like his subject, always seems intent on proving that about which he has already made up his mind.

Weisheipl is also the best source for the life and career of Albertus Magnus. *Albertus Magnus and the Sciences: Commemorative Essays, 1980*, which Weisheipl edited, is particularly useful in gaining perspective on the incredible range of Albert's work. Another helpful if somewhat reverential biography is *Albert the Great*, by another Dominican, the aptly named S. M. Albert.

For a tapestry of thirteenth-century life, there is no better place to look than to the works of the chroniclers of the period, particularly Matthew Paris and Thomas of Eccleston. Their writings are rich, written with sweep and wit, and filled with anecdotes that bring the period to life. Gordon Leff's *Medieval Thought* is an excellent source for the intellectual trends of the period, and Norman F. Cantor is always useful for a general backdrop. Volume IV of Will Durant's epic *Story of Civilization, The Age of Faith,* and *The Catholic Encyclopedia*, are both excellent sources as well. For the development of the universities, Stephen Ferruolo's *The Origins of the University* is a layered and detailed study, as is Leff's *Paris and Oxford Universities in the Thirteenth and Fourteenth Centuries*. Rosalind Brooke's *Early Franciscan Government: Elias to Bonaventure* paints a helpful picture of the development and political intrigues of the Friars Minor.

There are a number of excellent studies of Avicenna and Averroës, including Roger Arnaldaz's *Averroës: A Rationalist in Islam*, and *The Philosophy of Avicenna and Its Influences on Medieval Europe* by A. M. Goichon, and especially Majid Fakhry's excellent monograph *Averroës, Aquinas and the Rediscovery of Aristotle in Western Europe*. For Frederick, one need look no further than *Frederick the Second, 1194–1250*, Ernst Kantorowicz's brilliant and exhaustive study. *Emperor Frederick II* by David Geinstein is also useful.

For both John Dee and Francis Bacon, their own work provides not simply a professional record but also insights into the personalities of these two Elizabethan giants. Dee's diary is perhaps the best vehicle with which to appreciate the split of Roger Bacon's reputation into the dual streams of scientist and mystic.

Biographical accounts of the two men, particularly of Bacon, are less helpful and in general tend to be surprisingly one-sided.

As for the study of codes, code breaking, and the history of steganographics, there are no greater sources than *The Code Book* by Simon Singh and David Kahn's epic *The Codebreakers*. For an unusual and charming take, a reader will be well served to pick up *In Code* by a remarkable seventeen-year-old prodigy named Sarah Flannery. To learn about Walsingham, Dee, and Elizabethan intrigue, the best source is Richard Deacon's *History of the British Secret Service*, with Conyers Read's *Mr. Secretary Walsingham and the Policy of Queen Elizabeth* a close second. Although it is often considered dated, we found *The Follies of Science at the Court of Rudolph II, 1576–1612* by Henry Carrington Bolton profitable, as were Hans Holzer's *The Alchemist: The Secret Magical Life of Rudolf von Habsburg* and R. J. W. Evans's *Rudolph II and His World: A Study in Intellectual History, 1576–1612*.

Although there is no shortage of written material on the Voynich manuscript, Robert Brumbaugh's *The Most Mysterious Manuscript* and Mary D'Imperio's *Elegant Enigma* being both the most prominent and most useful, the best place to learn about the manuscript is on the Internet. There are dozens of Web sites devoted to the manuscript, from the dense and scholarly to the New Agey and fanciful. Though it is fun to come across a theory that the manuscript is an extraterrestrial artifact, the most useful Web sites are those of the international Voynich study group that was created in the early 1990s. Anyone who is interested should begin with René Zandbergen's site, www.voynich.nu, and go from there. The Beinecke has posted images of some of the pages from the manuscript, at http://highway49.library.yale.edu/photonegatives (cookies required—enter "Voynich" in search box).

# SELECTED BIBLIOGRAPHY

• • •

Albert, S. M., O. P. *Albert the Great*. Oxford: Blackfriars Publications, 1948.

Aquinas, St. Thomas. *Summa Contra Gentiles*. Edited by A. C. Pegis, et al. Notre Dame: University of Notre Dame Press, 1957–59.

———. *Summa Theologica*. 2nd edition. Edited by Father Laurence Shapcote. Chicago: Encyclopædia Britannica, 1990.

Aristotle. *The Complete Works: The Oxford Translation*. Translated by Jonathan Barnes. Princeton: Princeton University Press, 1995.

Arnaldaz, Roger. *Averroës: A Rationalist in Islam*. Notre Dame: University of Notre Dame Press, 2000.

Bacon, Francis. *The Advancement of Learning and Novum Organum*. London: Colonial Press, 1906.

———. *The Major Works*. Edited by Brian Vickers. Oxford: Oxford University Press, 1996.

Bacon, Roger. *Opera quaedam hactenus inedita*. Edited by James. S. Brewer. London: Longman, 1859.

———. *The* Opus Majus *of Roger Bacon*. Translated by Robert Belle Burke. Philadelphia: University of Pennsylvania Press, 1928.

———. *Part of the* Opus Tertium *of Roger Bacon*. Edited by A. G. Little. Aberdeen: University Press, 1912.

———. *Roger Bacon's Letter Concerning the Marvelous Power of Art and of Nature and Concerning the Nullity of Magic*. Edited by Tenney L. Davis. Kila, Mont.: Kessinger, 1997.

———. *Roger Bacon's Philosophy of Nature: A Critical Edition . . . of* De multiplicatione specierum *and* De speculis comburentibus. Edited and translated by David S. Lindberg. South Bend, Ind.: St. Augustine Press, 1998.

Bamford, James. *The Puzzle Palace: A Report on America's Most Secret Agency*. Boston: Houghton Mifflin, 1982.

Beamish, Tufton. *Battle Royal: A New Account of Simon de Montfort's Struggle Against King Henry III*. London: Frederick Muller, 1965.

Biller, Peter, and Barrie Dobson, eds. *The Medieval Church: Universities, Heresy and the Religious Life: Essays in Honour of Gordon Leff.* Great Britain: Ecclesiastical History Society, Boydell Press, 1999.

Birch, Thomas D. D., Secretary to the Royal Society. *The History of the Royal Society of London for Improving of Natural Knowledge, from Its First Rise.* Vols. 1–4. London: A. Millar in the Strand, 1754.

Bolton, Henry Carrington. *The Follies of Science at the Court of Rudolph II, 1576–1612.* Milwaukee: Pharmaceutical Review, 1904.

Bridges, John Henry. *The Life and Work of Roger Bacon: An Introduction to the* Opus Majus. London: Williams and Norgate, 1914.

Brooke, Rosalind B. *Early Franciscan Government: Elias to Bonaventure.* Cambridge: Cambridge University Press, 1959.

Brumbaugh, Robert S., ed. *The Most Mysterious Manuscript: The Voynich "Roger Bacon" Cipher Manuscript.* Carbondale, Il.: Southern Illinois University Press, 1978.

Cantor, Norman F. *The Civilization of the Middle Ages.* New York: HarperCollins, 1993.

*Catholic Encyclopedia.* New York: D. Appleton, 1914.

Crombie, A. C. *Robert Grosseteste and the Origins of Experimental Science, 1100–1700.* Oxford: Oxford University Press, 1953.

Deacon, Richard. *A History of the British Secret Service.* New York: Taplinger, 1969.

Dee, John. *The Mathematicall Praeface to the Elements of Geometrie of Euclid of Megara.* With an introduction by Allen G. Debus. New York: Science History Publications, 1975.

D'Imperio, Mary. *The Voynich Manuscript: An Elegant Enigma.* Washington, D.C.: National Security Agency, 1976.

Durant, Will. *The Story of Civilization.* New York: Simon and Schuster, 1954–75.

Easton, Stewart C. *Roger Bacon and His Search for a Universal Science.* New York: Columbia University Press, 1952.

Eccleston, Brother Thomas. *The Coming of the Friars Minor to England and Germany.* Translated by E. Gurney Salter. London: J. M. Dent and Sons, 1926.

Einstein, David G. *Emperor Frederick II.* New York: Philosophical Library, 1949.

Eiseley, Loren. *The Man Who Saw Through Time.* New York: Charles Scribner's Sons, 1973.

Evans, R. J. W. *Rudolph II and His World: A Study in Intellectual History, 1576–1612.* Oxford: Clarendon Press, 1973.

Fakhry, Majid. *Averroës, Aquinas and the Rediscovery of Aristotle in Western Europe*. Washington, D.C.: Center for Muslim-Christian Understanding, Georgetown University, 1997.

Ferruolo, Stephen C. *The Origins of the University: The Schools of Paris and Their Critics, 1100–1215*. Stanford: Stanford University Press, 1985.

Flannery, Sarah. *In Code*. New York: Workman, 2001.

French, Peter. *John Dee: The World of an Elizabethan Magus*. London: Ark Paperbacks, 1987.

Godwin, Joscelyn. *Athanasius Kircher: A Renaissance Man and the Quest for Lost Knowledge*. London: Thames and Hudson, 1979.

Goichon, A. M. *The Philosophy of Avicenna and Its Influences on Medieval Europe*. Delhi: Motilal Banarsidass, 1969.

Hackett, Jeremiah, ed. *Roger Bacon & the Sciences: Commemorative Essays*. Leiden: Brill, 1997.

Halliwell, James Orchard, ed. *The Private Diary of Dr. John Dee, and the Catalogue of His Library of Manuscripts*. London: John Bowyer Nichols and Son, 1892.

Haskins, Charles Homer. *Studies in Medieval Culture*. New York: Frederick Ungar, 1929.

Hawking, Stephen. *A Brief History of Time*. New York: Bantam Books, 1988.

Holland, Cecelia. *Antichrist: A Novel of the Emperor Frederick II*. New York: Atheneum, 1970.

Holzer, Hans. *The Alchemist: The Secret Magical Life of Rudolph von Habsburg*. New York: Stein and Day, 1974.

Johnson, Paul. *Elizabeth I*. New York: Holt, Rinehart and Winston, 1974.

Jordan, William Chester. *Louis IX and the Challenge of the Crusade: A Study in Rulership*. Princeton: Princeton University Press, 1979.

Kahn, David. *The Codebreakers: The Story of Secret Writing*. New York: Charles Scribner & Sons, 1996.

Kantorowicz, Ernst. *Frederick the Second, 1194–1250*. Translated by E. O. Lorimer. New York: Frederick Ungar, 1931.

Kaplan, Robert. *The Nothing That Is: A Natural History of Zero*. Oxford: Oxford University Press, 1999.

Labarge, Margaret Wade. *Saint Louis: Louis IX, Most Christian King of France*. Boston: Little, Brown, 1968.

Leff, Gordon. *Medieval Thought from St. Augustine to Ockham*. London: Merlin Press, 1959.

———. *Paris and Oxford Universities in the Thirteenth and Fourteenth Centuries: An Institutional and Intellectual History*. Huntington, N.Y.: Robert E. Krieger, 1975.

Little, A. G., ed. *Commemorative Essays*. Oxford: Oxford University Press, 1914.

———. *Franciscan Papers, Lists, and Documents*. Manchester: Manchester University Press, 1943.

McInerny, Ralph. *Aquinas Against the Averroists: On There Being Only One Intellect*. West Lafayette, Ind: Purdue University Press, 1993.

———. *A First Glance at St. Thomas Aquinas: A Handbook for Peeping Thomists*. Notre Dame: University of Notre Dame Press, 1990.

———. *St. Thomas Aquinas*. Notre Dame: University of Notre Dame Press, 1982.

Morris, Jan. *The Oxford Book of Oxford*. Oxford: Oxford University Press, 1978.

Newbold, William R. *The Cipher of Roger Bacon*. Edited by Roland G. Kent. Philadelphia: University of Pennsylvania Press, 1928.

———. *The Voynich Roger Bacon Manuscript*. Transactions of the College of Physicians of Philadelphia, 3rd ser., vol. 43. Philadelphia, 1921.

Osler, Margaret J., ed. *Rethinking the Scientific Revolution*. Cambridge: Cambridge University Press, 2000.

Paris, Matthew. *Matthew Paris's English History*. Translated by Rev. G. A. Giles. London: Bohn, 1854.

Plato. *The Dialogues of Plato*. Translated by Benajmin Jowett. Oxford: Oxford University Press, 1924.

Read, Conyers. *Mr. Secretary Walsingham and the Policy of Queen Elizabeth*. Cambridge: Harvard University Press, 1925.

Reeds, James. "William F. Friedman's Transcription of the Voynich Manuscript." *Cryptologia* 19 (1995):

Rosen, Edward. *Copernicus and the Scientific Revolution*. Malaber, Fla.: Robert E. Krieger, 1984.

Rowland, Ingrid D. *The Ecstatic Journey: Athanasius Kircher in Baroque Rome*. Chicago: University of Chicago Library, 2000.

Schama, Simon. *A History of Britain: At the Edge of the World? 3500 BC–1603 AD*. New York: Talk Miramax Books, Hyperion, 2000.

Sighart, Doctor Joachim. *Albert the Great of the Order of the Friar-Preachers: His Life and Scholastic Labours*. Translated by the Reverend Fr. T. A. Dixon. London, R. Washbourne, 1876.

Simeti, Mary Taylor. *Travels with a Medieval Queen*. New York: Farrar Straus Giroux, 2001.

Singh, Simon. *The Code Book*. New York: Doubleday, 1999.

Southern, Richard W. *Robert Grosseteste: The Growth of an English Mind in Medieval Europe*. Oxford: Clarendon Press, 1986.

Sprat, Thomas. *History of the Royal Society*. Edited by Jackson I. Cope and Harold Whitmore James. St. Louis: Washington University Studies, 1958.

Stephen, Leslie, et al. *Dictionary of National Biography*. Oxford: Oxford University Press, 1967–68.

Stimson, Dorothy. *Scientists and Amateurs: A History of the Royal Society*. London: Abelard-Schuman, 1946.

Thomson, Garret, and Marshall Missner. *On Aristotle*. Belmont, Calif.: Wadsworth, 2000.

Thorndike, Lynn. *A History of Magic and Experimental Science*. New York: Columbia University Press, 1923–58.

Tyack, Geoffrey. *Oxford: An Architectural Guide*. Oxford: Oxford University Press, 1998.

Voynich, Wilfrid M. *A Preliminary Sketch of the History of the Roger Bacon Cipher Manuscript*. Transactions of the College of Physicians of Philadelphia, 3rd ser., vol 43. Philadelphia, 1921.

Warren, W. L. *King John*. New York: W. W. Norton, 1961.

Weisheipl, James. *Friar Thomas D'Aquino: His Life, Thought and Works*. Garden City, N.Y.: Doubleday, 1974.

Weisheipl, James, ed. *Albertus Magnus and the Sciences: Commemorative Essays, 1980*. Toronto: Pontifical Institute of Mediaeval Studies, 1980.

Weld, Charles R. *A History of the Royal Society*. London: John R. Parker, 1848.

White, Andrew Dixon. *A History of the Warfare Between Science and Theology in Christendom*. New York: Appleton, 1896.

Wilkinson, B. *The Later Middle Ages in England, 1216–1485*. London: Longman, 1969.

Woolley, Benjamin. *The Queen's Conjurer: The Science and Magic of Dr. John Dee, Adviser to Queen Elizabeth I*. New York: Henry Holt, 2001.

# INDEX

Italicized page numbers indicate illustrations

# INDEX

## ABOUT THE AUTHORS

© WENDY KAHN

LAWRENCE AND NANCY GOLDSTONE are a husband-and-wife writing team and authors of *Out of the Flames*, a BookSense 76 Selection. They have also written three books on their book-collecting pursuits: *Used and Rare, Slightly Chipped*, and *Warmly Inscribed*.